John Oprea
101 Curl Dr.
Columbus, Ohio
43210

Other Monographs in this Series

No. 1. Irving Kaplansky: *Algebraic and analytic aspects of operator algebras.*

3. Lawrence Markus: *Lectures in Differentiable Dynamics.*

4. H. S. M. Coxeter: *Twisted honeycombs.*

5. George W. Whitehead: *Recent advances in homotopy theory.*

6. Walter Rudin: *Lectures on the edge-of-the wedge theorem.*

7. Yozo Matsushima: *Holomorphic vector fields on compact Kähler manifolds.*

Conference Board of the Mathematical Sciences

REGIONAL CONFERENCE SERIES IN MATHEMATICS

supported by the

National Science Foundation

Number 2

LECTURE NOTES ON NILPOTENT GROUPS

by

Gilbert Baumslag

Published for the
Conference Board of the Mathematical Sciences
by the
American Mathematical Society
Providence, Rhode Island

Expository Lectures
from the CBMS Regional Conference
held at the University of Texas, Austin, Texas
May 26–30, 1969

International Standard Book Number 0-8218-1650-0
Library of Congress Catalog Number 78-145636
AMS 1970 Subject Classifications: Primary 20 E 15;
Secondary 20 E 35, 22 E 25, 22 E 60.
Key words and phrases: Nilpotent group, algorithmic problems, residually finite, commutator calculus, lie and associative ring techniques, lie groups, matrix groups.

Printed in the United States of America

CONTENTS

Introduction

These lectures are concerned, in the main, with finitely generated nilpotent groups. The theory of these groups is rich and exciting.

There seem to be three main parts to this theory. The first of these deals with the so-called commutator calculus, which was initiated by Philip Hall in his fundamental paper [29]. As the name suggests it is concerned with manipulations of commutators and deductions of further relationships between commutators from basic identities. There does not seem to be any guiding principle in this calculus; consequently this aspect of the theory is for the most part rather difficult.

The second aspect of the theory is, in a sense, governed by a single principle. This principle may be likened to a well-known procedure in elementary number theory where one shows that a proposition about the integers holds modulo each prime p and thence for the integers themselves. This notion may be used, in particular, to prove certain results about finitely generated torsion-free nilpotent groups. The key fact is the following theorem of K. W. Gruenberg [25]: if G is a finitely generated torsion-free nilpotent group and p is any prime, then, given any element $g \in G$ $(g \neq 1)$, there is a normal subgroup N of G such that G/N is a finite p-group with $g \notin N$. Roughly speaking, the idea is to show that if a proposition about a finitely generated torsion-free nilpotent group holds for all its homomorphic images of prime-power order, then it holds also for the group itself.

The third part of nilpotent group theory stems in the main from the connection between lie groups and lie algebras; it was discussed first by A. I. Mal'cev in his beautiful paper [63]. The impact of this connection and the consequent connections between arithmetic and algebraic groups has only very recently emerged (see, for example, the very deep papers by L. Auslander [2], [3] and the paper by L. Auslander and G. Baumslag [4]). In a sense this is the most exciting, although it is in some ways the most limited, aspect of the whole theory. Here we shall develop *ab initio*, by using the approach of S. A. Jennings [47], as much of the necessary machinery as is needed for our discussion of the automorphism groups of finitely generated nilpotent groups.

Although I have chosen to divide the theory of finitely generated nilpotent groups into three parts, it should be pointed out that these parts are really very much interrelated, each complementing the others.

The program for these lectures is set forth in the table of contents. I have not divided the material up into the three parts described above because this would make for very awkward exposition. I have also not tried to provide a complete account of the theory here. The interested reader will find that a study of the bibliography included at the end of these lectures will allow him to go more deeply into those aspects of the theory which appeal to him most.

Acknowledgement

These notes are substantially the same as those prepared as an aid to the ten lectures on finitely generated nilpotent groups which I gave in Austin at the University of Texas during May of 1969. They have been slightly polished and numerous mistakes have been eradicated, mainly due to the diligence of John F. Ledlie to whom I would like to express my appreciation for his valuable help and assistance. I would also like to express my thanks to Nancy Singleton and Kathy Vigil for their patient deciphering of many rewritten, half legible pages which made up this manuscript. Finally I would like to thank the Mathematics Department (especially John R. Durbin) of the University of Texas for making it such a pleasure to spend a week in May in Austin.

Chapter 0. Basic notions and results

This brief section contains some material which is basic in the sequel. Most proofs in this section will either be very much abbreviated or omitted altogether. Complete proofs can either be furnished directly or can be extracted from the fundamental paper by P. Hall [29], or from the book by M. Hall, Jr. [27], or obtained from the references cited.

0.1. Let X be a property (or class) of groups. A finite normal series (i.e., each term in the series is a normal subgroup of the succeeding term)

$$1 = G_0 \leq G_1 \leq \cdots \leq G_l = G \tag{1}$$

of G is termed a *poly-X series for* G if $G_{i+1}/G_i \in X$ for $i = 0, \cdots, l-1$. A group G is termed *poly-X* if it has at least one poly-X series. The *length* of the series (1) is l.

Polyabelian groups are usually referred to as *solvable*.

The center of a group G is denoted by ζG. An invariant series (1) (i.e., G_i is a normal subgroup of G for $i = 1, \cdots, l$) is termed a *central series* if $G_{i+1}/G_i \leq \zeta(G/G_i)$ for $i = 0, \cdots, l-1$. G is *nilpotent* if it has at least one central series; the *class* of a nilpotent group is the least of the lengths of its central series.

Finitely generated nilpotent groups are polycyclic (this can be proved by observing that if G is a finitely generated nilpotent group of class c then $\lambda_c G$ is also finitely generated (see 0.2 below)).

An extension of a finitely presented group by another finitely presented group is finitely presented. Hence poly-finitely-presented coincides with finitely presented. In particular then, polycyclic groups are finitely presented.

A group G satisfies the *maximal condition* if it has no infinite strictly increasing chain of subgroups; this is equivalent to the condition that all subgroups of G are finitely generated. An extension of a group with maximal condition by another such group satisfies the maximal condition. Thus the poly-maximal-condition coincides with the maximal condition. In particular polycyclic groups satisfy the maximal condition.

Let p be any prime and let G be a nilpotent group. Then the elements of order a power-of-p constitute a normal subgroup of G, the Sylow p-subgroup of G.

If G is any group then G' denotes the commutator or derived group of G. A nilpotent group G is abelian if and only if $G/\zeta G$ is cyclic; hence if $a \in G$ and G is of class $c > 1$ then $gp(a, G')$ is of class at most $c - 1$. So G is cyclic if and only if G/G' is cyclic. It follows that if G is nilpotent and τG denotes the set of elements of finite order in G, then τG is the restricted direct product of the Sylow p-subgroups of G. We call τG the *torsion subgroup* of G; τG is a normal subgroup of G and $G/\tau G$ is torsion-free.

Since finitely generated nilpotent groups satisfy the maximal condition, a finitely generated nilpotent torsion group is finite.

The number of infinite cyclic factors in any polycyclic series for a polycyclic group G is an invariant, the so-called *torsion-free rank* of G (K. A. Hirsch [43]). However the presence of finite factors in such a series is more awkward to keep track of and difficult to categorize (K. A. Hirsch [43], J. F. Bowers [17]). Thus for example there exist torsion-free polycyclic groups which are not poly-infinite-cyclic although finitely generated torsion-free nilpotent groups are always poly-infinite-cyclic.

Suppose that G is finitely generated and nilpotent. Since τG is finite it has a composition series of length l, say. The pair (t, l) will be termed the *general rank of* G, where t is the torsion-free rank of G. It follows from our initial comments that the general rank of G is well defined.

It is easy to prove, by making use of the "basis theorem" for finitely generated abelian groups, the following

Theorem 0.1. *Let G be a finitely generated nilpotent group and let N be a non-trivial normal subgroup of G. If (t, l) is the general rank of G and (t', l') is the general rank of G/N, then either $t' < t$ or if $t' = t$ then $l' < l$.*

A group is termed *hopfian* if $G/N \cong G$ implies $N = 1$. It follows immediately from Theorem 0.1 that

Corollary 0.11. *Finitely generated nilpotent groups are hopfian.*

0.2. Let $x_1, x_2, \cdots, x_n$ be elements of a group G. We define

$$[x_1, x_2] = x_1^{-1}x_2^{-1}x_1x_2,\ x_1^{x_2} = x_2^{-1}x_1x_2,$$

and recursively for $n > 2$,

$$[x_1, \cdots, x_n] = [[x_1, \cdots, x_{n-1}], x_n].$$

The following identities are now standard—they are due to P. Hall.

Theorem 0.2. *If G is any group and $x, y, z \in G$, then*

(i) $[xy, z] = [x, z]^y[y, z]$, $[x, yz] = [x, z][x, y]^z$;

(ii) $[x, y^{-1}, z]^y[y, z^{-1}, x]^z[z, x^{-1}, y]^x = 1$.

This notation and the results (i), (ii) (especially the remarkable identity (ii)) constitute the starting point of the so-called *commutator calculus*.

If G is any group and $H_1, H_2, \cdots, H_n$ are any subsets of G, then we define

$$[H_1, H_2] = gp([h_1, h_2] | h_1 \in H_1, h_2 \in H_2),$$

and recursively for $n > 2$

$$[H_1, H_2, \cdots, H_n] = [[H_1, \cdots, H_{n-1}], H_n].$$

The *lower central series* $G = \lambda_1 G \geq \lambda_2 G \geq \cdots$ of a group G is defined by $\lambda_i G = [G, G, \cdots, G]$ (i terms altogether). Observe that G is nilpotent of class $c \geq 1$ if and only if $\lambda_c G \neq 1$ and $\lambda_{c+1} G = 1$. The derived series $G = \delta_0 G \geq \delta_1 G \geq \cdots$ is defined recursively by $\delta_{i+1} G = (\delta_i G)'$.

An easy consequence of Theorem 0.2 (ii) is the following "three subgroup theorem" of P. Hall.

Theorem 0.3. *If A, B and C are normal subgroups of a group G then*

$$[A, B, C] \leq [B, C, A][C, A, B].$$

The *upper central series*

$$1 = v_0 G \leq v_1 G \leq \cdots$$

of a group G is defined recursively by

$$v_{i+1} G / v_i G = \zeta(G / v_i G).$$

It follows then by induction using Theorem 0.3 that

Corollary 0.31.

$$[\lambda_i G, \lambda_j G] \leq \lambda_{i+j} G, \; for \; i, j \geq 1.$$

Corollary 0.32.

$$\delta_i G \leq \lambda_{2^i} G, \; for \; i = 1, 2, \cdots.$$

Corollary 0.33.

$$[v_i G, \lambda_i G] = 1, \; for \; i = 1, 2, \cdots.$$

0.3. The properties of the center of a nilpotent group are often reflected in the entire group. One such result is the following (P. Hall [35]):

Lemma 0.1. *A finitely generated nilpotent group G is finite if and only if its center is.*

Proof. If the class c of G is 1 there is nothing to prove.

Suppose $c > 1$. If $v_1 G$ has order n, if $a \in v_2 G$ and $x \in G$ then

$$[x, a^n] = [x, a]^n = 1.$$

So $a^n \in v_1 G$ and $v_2 G/v_1 G$ is finite since $G/v_1 G$ satisfies the maximal condition. Inductively $G/v_1 G$ is finite and so G is also.

0.4. If X is any class (or property) of groups, then $\mathbf{R}X$ denotes the class of groups which are residually in X, i.e., the class of subdirect products of X-groups. Thus a group G belongs to $\mathbf{R}X$ if for each $g \in G$ $(g \neq 1)$ there exists a normal subgroup N_g of G such that

$$G/N_g \in X \text{ and } g \notin N_g.$$

We denote the class of all finite groups by $\mathfrak{F}$, the class of all finite p-groups (p a fixed prime) by $\mathfrak{F}_p$ and the class of all nilpotent groups by $\mathfrak{N}$.

0.5. Let π be a set of primes. An integer is termed a *π-number* if its prime divisors lie in π. A group is termed a *π-group* if every element has (finite) order a π-number. A group is said to be *π-free* if none of its elements has order a π-number.

A group G is termed a *$\mathfrak{U}_\pi$-group* if

$$g^p = h^p \text{ implies } g = h \ (g, h \in G)$$

whenever $p \in \pi$. A $\mathfrak{U}_\pi$-group is obviously π-free. A subgroup H of a group G is *π-isolated in* G if

$$g^p \in H \text{ implies } g \in H \ (g \in G)$$

whenever $p \in \pi$. It is clear that a normal subgroup N of G is π-isolated in G if and only if G/N is π-free.

Theorem 0.4 (**P. G. Kontorovic**). *The centralizer of any subset S of a $\mathfrak{U}_\pi$-group G is π-isolated in G.*

Proof. Suppose x^p centralizes S, where $p \in \pi$. Then for each $s \in S$,

$$x^p = s^{-1} x^p s = (s^{-1} x s)^p.$$

As $G \in \mathfrak{U}_\pi$, this implies $x = s^{-1} x s$; therefore x centralizes S.

Corollary 0.41. *If $G \in \mathfrak{U}_\pi$ then $G/\zeta G \in \mathfrak{U}_\pi$.*

Proof. Denote ζG by Z and suppose $(xZ)^p = (yZ)^p$ $(x, y \in G, p \in \pi)$. Then $x^p = y^p z$ for some $z \in Z$. It follows that

$$(y^{-1} x y)^p = y^{-1} x^p y = y^{-1} y^p z y = y^p z = x^p.$$

So $y^{-1} x y = x$. Therefore $(y^{-1} x)^p = y^{-p} x^p \in Z$, and hence $y^{-1} x \in Z$ because Z is π-isolated in G by Theorem 0.4. This completes the proof.

Chapter 1. Algorithmic problems for finitely generated nilpotent groups

Three problems--the word, conjugacy and isomorphism problems-were raised by M. Dehn [19] early in this century. Here we discuss there problems in the context of finitely generated nilpotent groups (see A. W. Mostowski [70], [71]).

1.1. We consider first the word problem (see M. O. Rabin [78] for a good discussion of this and related problems). To this end we introduce the notation

$$G = (x_1, \cdots, x_m; r_1, \cdots, r_n)$$

to express the fact that G is generated by $x_1, \cdots, x_m$ and completely defined in terms of these generators by the relations $r_1 = \cdots = r_n = 1$ (see e.g., W. Magnus, A. Karrass and D. Solitar [61, Chapter 1]).

The guiding light in this discussion is the following simple theorem of J. C. C. McKinsey [66] (see also V. H. Dyson [21]).

Theorem 1.1. *A finitely presented residually finite group has a solvable word problem.*

Proof. Let F be a free group on $x_1, \cdots, x_m$, R the normal subgroup of F generated by $r_1, \cdots, r_n$ and suppose $F/R \in \mathbf{R}\mathfrak{F}$. We have to provide an algorithm for deciding whether an element $f \in F$ actually lies in R or not. We simultaneously carry out two procedures. On the one hand we begin an effective enumeration of the elements of R. On the other hand we begin to effectively list all (multiplication tables of) the finite factor groups of F/R. Since $F/R \in \mathbf{R}\mathfrak{F}$, either f turns up in the enumeration of the elements of R or else a nonidentity image of f will show up in the list of finite factor groups of F/R. This solves the word problem in F/R.

Since, as we observed earlier, finitely generated nilpotent groups are finitely related, it is enough to prove that polycyclic groups are $\mathbf{R}\mathfrak{F}$ in order to settle positively the word problem for finitely generated nilpotent groups. We prove somewhat more, viz.

Theorem 1.2. *A cyclic extension of a finitely generated residually finite group is residually finite.*

Proof. Let H be a finitely generated residually finite group and suppose H is a normal subgroup of G. Suppose, furthermore, that

$$G = gp(a, H);$$

so G/H is cyclic on aH. We shall prove that $G \in \mathbf{R}\mathfrak{F}$.

Let $g \in G$ $(g \neq 1)$. If $g \notin H$ we can find a normal subgroup N_g of G such that $g \notin N_g$ and $G/N_g \in \mathfrak{F}$ since $G/H \in \mathbf{R}\mathfrak{F}$. So we may assume $g \in H$. Now H is a finitely generated $\mathbf{R}\mathfrak{F}$ group. Since a finitely generated group has only a finite number of subgroups of a given finite index (see M. Hall, Jr. [28]), there is a characteristic subgroup C of H of finite index such that $g \notin C$. If G/H is finite, we can take $N_g = C$; so we may assume G/H is infinite cyclic. Put $\bar{G} = G/C$, $\bar{H} = H/C$, $\bar{g} = gC$ and $\bar{a} = aC$. Let $\bar{D}$ be the centralizer of $\bar{H}$ in $\bar{G}$. Since $\bar{H}$ is finite, $\bar{D}$ is of finite index in $\bar{G}$. It follows that some power $\bar{a}^m$ of $\bar{a}$ $(m \neq 0)$ centralizes $\bar{H}$. Clearly, since a is of infinite order modulo H,

$$\bar{g} \notin gp(\bar{a}^m);$$

further $gp(\bar{a}^m)$ is normal in $\bar{G}$ and $\bar{G}/gp(\bar{a}^m)$ is finite. This group $\bar{G}/gp(\bar{a}^m)$ is a finite factor group of $\bar{G}$ in which the image of g is not the identity. This completes the proof of the theorem.

An immediate consequence of Theorem 1.2 is the

Corollary 1.21. *Polycyclic groups are* $\mathbf{R}\mathfrak{F}$.

Corollary 1.21 is due to K. A. Hirsch [43]. Thus all polycyclic groups (which, of course, include all finitely generated nilpotent groups) have solvable word problem.

Incidentally, it is not difficult to derive this latter result more directly by constructing a polycyclic series from a given (presentation of a) polycyclic group.

1.2. Our approach to the word problem has been a little contrived; as we remarked the desired result could have been obtained by a direct approach. A more genuine use of residual finiteness in solving an algorithmic problem occurs if we consider the conjugacy problem for finitely generated nilpotent groups. The key fact is the following theorem of N. Blackburn [14] (the proof given here seems somewhat simpler than Blackburn's original proof).

Theorem 1.3. *Let G be a finitely generated nilpotent group. Then two elements of G are conjugate if and only if they are conjugate in every finite factor group of G.*

Proof. To indicate that u and v are conjugate elements of a group we shall write $u \sim v$; otherwise we write $u \not\sim v$.

The proof is by induction on the torsion-free rank r of G. If $r = 0$, G is finite and there is nothing to prove.

Thus we suppose $r > 0$. Let $g, h \in G$ be elements which are conjugate in every finite factor group of G. Suppose that

$$g \not\sim h;$$

we shall show that this leads to a contradiction.

By Lemma 0.1, ζG must contain an element a, say, of infinite order. We put

$$H_i = gp(a^{i!}), \quad i = 1, 2, \cdots.$$

It follows from Theorem 0.1 that the torsion-free rank of G/H_i is less than r (in fact it is $r-1$). If $gH_i \not\sim hH_i$ for some i, then inductively g and h are not conjugate in some finite factor group of G. So

$$gH_i \sim hH_i, \quad i = 1, 2, \cdots.$$

In particular then we can find an integer m ($\neq 0$) and an element $x \in G$ such that

$$h^x = ga^m.$$

It follows that

$$g \sim ga^m \text{ modulo } H_i, \text{ for } i = 1, 2, \cdots.$$

In other words, for each $i = 1, 2, \cdots$, we can find an element $y_i \in G$ and an integer s_i such that

$$y_i^{-1} g y_i = ga^m \cdot (a^{i!})^{s_i}. \tag{1}$$

Put

$$L = gp(y_1, y_2, \cdots, g, a).$$

Clearly $a \in \zeta L$ since $a \in \zeta G$. It follows from (1) that the mapping

$$\phi\colon l \longrightarrow [l, g], \; l \in L,$$

is a homomorphism of L into $gp(a)$. Thus if K is the kernel of ϕ, that is, the centralizer of g in L, then L/K is cyclic. So

$$L = gp(K, b),$$

for a suitable choice of $b \in L$. We may assume, on replacing b by b^{-1} if necessary, that

$$[b, g] = a^\alpha, \text{ with } \alpha \geq 0.$$

Further $\alpha \neq 0$, for $\alpha = 0$ implies $g \in \zeta L$ which invalidates (1).

Observe that the conjugates of g in L are simply the elements

$$g, \; ga^{\pm\alpha}, \; ga^{\pm 2\alpha}, \cdots.$$

Now since g and ga^m are not conjugate in G, α does not divide m. However, since $\alpha > 0$, we have from (1),

$$y_\alpha^{-1} g y_\alpha = g a^m (a^{\alpha!})^{s_\alpha}.$$

So $m + s_\alpha \alpha! = \lambda\alpha$, for a suitable choice of the integer λ. Hence α divides m. This contradiction completes the proof of Theorem 1.3.

It follows from Theorem 1.3, in much the same way as the word problem is solved for finitely presented $\mathbf{R}\mathfrak{F}$ groups, that

Corollary 1.31. *The conjugacy problem is solvable for finitely generated nilpotent groups.*

This corollary automatically raises the question: Does every polycyclic group have a solvable conjugacy problem?

In this direction, M. I. Kargapolov [49] has proved that every supersolvable group G has the property that two elements are conjugate in G if and only if they are conjugate in every finite factor group of G. (A *supersolvable* group is one possessing an invariant polycyclic series.) Very recently E. Formanek has proved the corresponding result for polycyclic groups.

The third problem raised by Dehn—the isomorphism problem—is probably the most interesting and also the most difficult of all. One might hope that for finitely generated nilpotent groups it is possible to effectively obtain a set of reasonable invariants which uniquely determine these groups. Very little information of any kind has been obtained however. It seems fairly easy to show that it is sufficient to deal only with the torsion-free case. One obvious approach, prompted by our discussion above, is to try to classify finitely generated torsion-free nilpotent groups by their finite homomorphic images. This however does not work since both V. N. Remeslennikov [79] and, more recently, Graham Higman have constructed nonisomorphic, finitely generated torsion-free nilpotent groups with the same finite homomorphic images.

Remeslennikov's examples are the more easily described. Thus let F be a free nilpotent group of class 4 on x and y, i.e., F is an isomorphic image of $H/\lambda_5 H$ where H is a free group on two generators a and b with x and y mapping under some isomorphism to $a\lambda_5 H$ and $b\lambda_5 H$ respectively. Let

$$S = gp([y, x, x, x]^2 [y, x, x, y][y, x, y, y]^3)$$

and

$$T = gp([y, x, x, x][y, x, x, y][y, x, y, y]^6).$$

Then it turns out that F/S and F/T have the same finite images but are not isomorphic.

Let us denote by $\mathbf{F}(G)$ the isomorphism classes of finite homomorphic images of the group G. It follows from our previous remarks that if A and B are finitely generated torsion-free nilpotent groups then $\mathbf{F}(A) = \mathbf{F}(B)$ does not always imply $A \cong B$. A. Borel (unpublished) has, however, shown that if A is any finitely generated torsion-free nilpotent group then the class $b(A)$ of those finitely generated torsion-free nilpotent groups B such that $\mathbf{F}(B) = \mathbf{F}(A)$ *is* bounded, in a sense. In order to explain let us term two groups X and Y *commensurable* if they contain subgroups X_1 and Y_1 respectively, of finite index, with $X_1 \cong Y_1$. Commensurability is an equivalence relation; thus we may speak of commensurability classes. Borel's "boundedness" result may then be formulated more precisely by: $b(A)$ decomposes into finitely many commensurability classes. Borel's proof makes use of [16]. It should be pointed out that G. Higman's examples show that even if $\mathbf{F}(A) = \mathbf{F}(B)$ A and B need not be commensurable, although F. Pickel has proved F/S and F/T *are* commensurable. More important, Pickel has proved that $b(A)$ actually decomposes into finitely many *isomorphism* classes for every finitely generated torsion-free nilpotent group A.

Chapter 2. Residual properties and some applications

In this section we shall show how the residual finiteness of finitely generated nilpotent groups can be exploited.

2.1. In a finitely generated abelian group the torsion subgroup is a direct factor. This is also true, in a sense, for finitely generated nilpotent groups.

Theorem 2.1 (K. A. Hirsch [43]). *Let G be a finitely generated nilpotent group. Then G can be embedded as a subgroup of finite index in a direct product $D = A \times B$, where A is finite and B is torsion-free.*

Proof. Let $T = \tau G$. Then T is finite. So, using $G \in \mathbf{R}\mathfrak{F}$, there is a normal subgroup N of G of finite index such that

$$N \cap T = 1.$$

Put $A = G/N$, $B = G/T$. Then A is finite and B is torsion-free. Since $N \cap T = 1$ the mapping

$$\phi : g \longrightarrow (gN, gT), \; g \in G,$$

of G into $D = A \times B$ is a monomorphism. Moreover $N\phi$ is a subgroup of finite index in B and hence $G\phi$ if of finite index in D. This completes the proof.

Theorem 2.1 allows us, in some ways, to divide the study of finitely generated nilpotent groups up into the study of finite nilpotent groups and finitely generated torsion-free nilpotent groups. As we saw in Chapter 0, finite nilpotent groups are simply the direct products of their Sylow p-subgroups. So the study of finite p-groups is relevant to the theory of finitely generated nilpotent groups. But this aspect of nilpotent groups is not our main concern here and will be disregarded, for the most part.

2.2. We know that polycyclic groups are $\mathbf{R}\mathfrak{F}$ (Corollary 1.21). A considerably stronger statement is true for finitely generated torsion-free nilpotent groups.

Theorem 2.2 (K. W. Gruenberg [25]). *Let G be a finitely generated torsion-free nilpotent group. If p is any given prime then $G \in \mathbf{R}\mathfrak{F}_p$.*

To enable us to make the inductive step in our proof of Theorem 2.2, we need

Lemma 2.1 (A. I. Mal'cev [64]). *If G is a π-free nilpotent group then $G \in \mathfrak{U}_\pi$.*

Proof. We proceed by induction on the class c of G. If $c = 1$ the result is immediate.

Suppose $c > 1$ and that $g, h \in G$, $p \in \pi$ are such that $g^p = h^p$. Then

$$g^p = h^p = h^{-1}h^ph = h^{-1}g^ph = (h^{-1}gh)^p.$$

But as we noted earlier, $gp(g, G')$ is of class at most $c-1$. Therefore $gp(g, G') \in \mathfrak{U}_\pi$. Now g and $h^{-1}gh \in gp(g, G')$. So $g^p = (h^{-1}gh)^p$ implies $g = h^{-1}gh$. Hence $(gh^{-1})^p = 1$. But G is π-free and $p \in \pi$. So $gh^{-1} = 1$ and $g = h$.

An immediate consequence of Lemma 2.1 and Corollary 0.41 is the

Corollary 2.11. *If G is a π-free nilpotent group then $G/\zeta G$ is π-free.*

We are now in a position to prove Theorem 2.2.

Proof of Theorem 2.2. Since infinite cyclic groups are $\mathbf{R}\mathfrak{F}_p$, so are finitely generated torsion-free abelian groups, and we may proceed to prove Theorem 2.2 by induction on the class c of G.

Suppose $c > 1$ and let $1 \neq g \in G$; we must find a normal subgroup N_g of G of p-power index, with $g \notin N_g$. Now by Corollary 2.11, $G/\zeta G$ is again torsion-free and it is of class at most $c-1$; inductively $G/\zeta G \in \mathbf{R}\mathfrak{F}_p$, so the case $g \notin \zeta G$ presents no problem. Thus we may assume $g \in \zeta G$. Choose a subgroup $L \leq \zeta G$ such that $g \notin L$ and $\zeta G/L \in \mathfrak{F}_p$. Since G satisfies the maximal condition, among all the normal subgroups of G which contain L and avoid g, there exists a maximal one M, say. We claim that $G/M \in \mathfrak{F}_p$. For G/M is a finitely generated nilpotent group; hence, by Corollary 1.21, $G/M \in \mathbf{R}\mathfrak{F}$. But gM is a nontrivial element of G/M which lies inside every nontrivial normal subgroup. The only possibility is that G/M is finite. Then G/M, being a finite nilpotent group, is the direct product of its Sylow subgroups. By the maximality of M, there can be only one such Sylow subgroup, so that $|G/M| = q^\alpha$ for some prime q. To show finally that $q = p$, note that G/M has the nontrivial subgroup $\zeta G \cdot M/M \cong \zeta G/\zeta G \cap M$ which, being a factor group of $\zeta G/L$ must be a finite p-group. This completes the proof of the theorem.

Actually Gruenberg proves a little more (and indeed this follows from the argument above) namely: if π is the set of prime divisors of the order of the torsion subgroup of a finitely generated nilpotent group G, then $G \in \mathbf{R}X$, where $X = \bigcup_{p\in\pi} \mathfrak{F}_p$.

The conclusion of Theorem 2.2 does not hold for polycyclic groups. Indeed if a polycyclic group is $\mathbf{R}\mathfrak{F}_p$ for infinitely many primes p, then it is nilpotent.* However, A. Learner [54] has proved that if G is a polycyclic group then there exists a finite set of primes π such that G is residually a finite π-group.

2.3. Theorem 2.2 has a number of applications which involve the existence and uniqueness of roots in groups (see [7]). In order to explain we need some notation: $\mathfrak{U}$ is the class of groups in which extraction of roots is unique when possible; $\mathfrak{E}$ is the class of groups in which extraction of roots is always possible; $\mathfrak{D}$ is the class of

---

* K. Seksendaev, *On the theory of polycyclic groups,* Algebra i Logika Sem. 4 (1965), no. 3, 79–83. (Russian) MR 33 #5735.

groups in which extraction of roots is always uniquely possible (thus $\mathfrak{D} = \mathfrak{U} \cap \mathfrak{E}$).

The relevance of these notions to periodic groups is provided by the following fact.

Lemma 2.2. *Let π be a set of primes, let G be a π-group and let p be a prime outside π. Then pth roots exist and are unique in G.*

Proof. Let g in G have order n. Since p and n are coprime, we can find integers α and β such that $\alpha p + \beta n = 1$. Then

$$g = g^1 = g^{\alpha p + \beta n} = (g^\alpha)^p.$$

So g^α is a pth root of g. To prove that it is the only pth root of g in G, suppose that $x^p = g$, $x \in G$. Then $(x^n)^p = g^n = 1$, so that $x^n = 1$, since G is a π-group. Therefore

$$x = x^1 = x^{\alpha p + \beta n} = x^{\alpha p} x^{\beta n} = g^\alpha.$$

This completes the proof.

One part of Lemma 2.2 immediately yields the

Proposition 2.1. *Let π and ρ be two nonempty disjoint sets of primes. If the group G is residually a π-group and also residually a ρ-group, then $G \in \mathfrak{U}$.*

Lemma 2.2 and Proposition 2.1 suggest the

Problem. Let π and ρ be two nonempty disjoint sets of primes. Suppose that the group G is residually a π-group and residually a ρ-group. Can G be embedded in a $\mathfrak{D}$-group?

Prompted by this problem we prove

Lemma 2.3. Let G be a finitely generated torsion-free nilpotent group of class c, let g be an element of G and let n be a positive integer. Then G can be embedded in a finitely generated torsion-free nilpotent group of class c in which g has an nth root.

Proof. Choose a prime p which does not divide n. Since $G \in \mathbf{R}\mathfrak{F}_p$ we can find normal subgroups $N_1, N_2, \cdots$ of G of p-power index such that $\bigcap_{i=1}^{\infty} N_i = 1$. Let P be the cartesian (i.e., unrestricted direct) product of the groups G/N_i:

$$P = \prod_{i=1}^{\infty} G/N_i.$$

Then the mapping ϕ: $g \longrightarrow (gN_i)$ is a monomorphism since the N_i intersect in 1; here (gN_i) stands for that element of P which has gN_i as its component in G/N_i. Now gN_i has an nth root, say h_i, in G/N_i (Lemma 2.2). Hence $h = (h_i)$ is an nth root of $g\phi$. Observe that P is nilpotent of class c since none of the G/N_i has class exceeding c. This implies that $H = gp(G\phi, h)$ is of class c. Finally, since

$G\phi \cap \tau H = 1$, it follows that G can be embedded in $H/\tau H$, a finitely generated torsion-free nilpotent group of class c, in which the image of g has an nth root.

2.4. It follows easily from Lemma 2.3 that every finitely generated torsion-free nilpotent group can be embedded in a nilpotent $\mathfrak{D}$-group. More generally by a standard technique in algebra (see, e.g., B. H. Neumann [72]) it follows that *every* torsion-free nilpotent group can be embedded in a nilpotent $\mathfrak{D}$-group. This result was proved first by A. I. Mal'cec [63]. We shall give a new direct proof of this theorem during the course of our discussion on the applications of residual properties. Here in this paragraph, we develop a formula due to P. Hall [29] which will greatly facilitate this discussion. This formula represents another phase of the commutator calculus.

We shall need the notion of an m-fold commutator in a group G. As usual a 1-fold commutator in G is simply an element of G. Inductively an m-fold commutator in G $(m > 1)$ is a commutator $[u, v]$ where u is an i-fold commutator in G, v is a j-fold commutator and $i + j = m$. So an m-fold commutator is a repeated commutator in m arguments which are sometimes referred to as the components of the m-fold commutator.

Suppose now that $p = y_1 y_2 \cdots y_n$ is a product of n elements $y_1, \cdots, y_n$ of a group G. Let R be the set of the first r natural numbers: $R = \{1, 2, \cdots, r\}$. Let λ be a mapping of the (indexed) set $y_1, \cdots, y_n$ into R. We term $y_i\lambda$ the *label* of y_i. The mapping λ will be fixed for most of this discussion.

Let S be any nonempty subset of R. Define X_S to be the set of all m-fold $(m \geq |S|)$ commutators c in $y_1, \cdots, y_n$ which satisfy the following conditions:

(i) the label of each component of c is in S,

(ii) each element of S is the label of some component of c.

Order the nonempty subsets of R first by cardinality and then lexicographically. The first fact we shall need is

Lemma 2.4. *$p = \Pi_{S \subseteq R} q_S$ where q_S is a product of elements in X_S and the $2^r - 1$ factors q_S occur in the order imposed on them (by the ordering of the subsets of R).*

Proof. Suppose y_v is the first y in p with label 1. If $v = 1$ we do nothing. If $v > 1$ we move y_v in front of y_{v-1} by observing that

$$y_{v-1} y_v = y_v y_{v-1} [y_{v-1}, y_v].$$

Of course $[y_{v-1}, y_v] \in X_S$ where $S = \{y_{v-1}\lambda, 1\}$. By repeating this process we can express p in the form

$$p = q_{\{1\}} \cdots q_{\{r\}} p_{21} \cdots p_{2l}$$

where p_{2j} is an element of an X_S where each S occurring has at least two labels in it it, for $j = 1, \cdots, l$. By iterating this procedure the required expression for p is obtained

Now let S be a given subset of R and put

$$p_S = y_{i_1} \cdots y_{i_t} \ (i_1 < i_2 < \cdots < i_t),$$

where $y_{i_1}, \cdots, y_{i_t}$ consist of all the y_j occurring in p whose labels lie in S. Thus p_S results from p by putting $y_k = 1$ if $y_k\lambda \notin S$. If these substitutions are made in the right-hand side of the expression obtained in Lemma 2.4 for p, we find that the effect is to make $q_T = 1$ whenever $T \not\subseteq S$ and to leave q_T unchanged if $T \subseteq S$. It follows therefore that we have proved

Lemma 2.5. *For each subset S of R,*

$$p_S = \prod_{T \subseteq S} q_T$$

where the factors q_T occur in their prescribed order.

The point of Lemma 2.5 is that it enables us to express each of the q_S in terms of the p_T, $T \subseteq S$, by recurrence. Thus we see that if $\alpha, \beta \in R$, $\alpha < \beta$, then

$$q_{\{\alpha\}} = p_{\{\alpha\}},$$

and

$$q_{\{\alpha,\beta\}} = q_{\{\alpha\}} q_{\{\beta\}} q_{\{\alpha,\beta\}};$$

so

$$q_{\{\alpha,\beta\}} = p_{\{\beta\}}^{-1} p_{\{\alpha\}}^{-1} p_{\{\alpha,\beta\}}.$$

We consider now the special case $p = y_1 \cdots y_{mr}$ where

$$y_1 = y_2 = \cdots = y_r = x_1, y_{r+1} = \cdots = y_{2r} = x_2, \cdots, y_{(m-1)r+1} = \cdots = y_{mr} = x_m.$$

Thus $p = x_1^r \cdots x_m^r$ where r is the number of elements in $R = \{1, 2, \cdots, r\}$. Define the labelling mapping λ of $y_1, \cdots, y_{mr}$ into R by putting $y_i\lambda = j$ $(1 \leq j \leq r)$ where j is defined by $j \equiv i(r)$.

Now let S be a subset of R. If $|S| = w$ then

$$p_S = x_1^w x_2^w \cdots x_m^w.$$

Observe that p_S depends here only on $|S|$, i.e., only on w. Hence it follows from our remarks above that q_S also depends only on $|S|$. Thus we may write here

$$q_S = \tau_w(x_1, \cdots, x_m) = \tau_w(\mathbf{x}), \text{ say.}$$

But now we may use Lemma 2.5 to actually express $x_1^w x_2^w \cdots x_m^w$. The net result, which is an immediate consequence of Lemma 2.5 and the preceding discussion, is the following theorem of P. Hall.

Theorem 2.3. *If w is any positive integer and if $x_1, \cdots, x_m$ are elements of a group then*

$$x_1^w x_2^w \cdots x_m^w = \tau_1(\mathbf{x})^w \tau_2(\mathbf{x})^{\binom{w}{2}} \cdots \tau_{w-1}(\mathbf{x})^w \tau_w(\mathbf{x})$$

where $\tau_i(\mathbf{x})$ is a product of j-fold commutators ($j \geq i$) in $x_1, \cdots, x_m$ for $i = 1, \cdots, w$.

Corollary 2.31. *Let G be a nilpotent group of class c and let p be any prime which is bigger than c. Then a product of p^nth powers of elements of G is again a p^nth power, for every $n \geq 1$.*

Proof. Suppose that $G = gp(x_1, \cdots, x_m)$ is of class c and p is a prime bigger than c. If $c = 1$ then

$$x_1^{p^n} \cdots x_m^{p^n} = (x_1 \cdots x_m)^{p^n}.$$

So we shall assume that the assertion has been proved for all groups of class less than c where $c > 1$. Now by Theorem 2.3

$$x_1^{p^n} x_2^{p^n} \cdots x_m^{p^n} = \tau_1(\mathbf{x})^{p^n} \tau_2(\mathbf{x})^{\binom{p^n}{2}} \cdots \tau_{p-1}(\mathbf{x})^{\binom{p^n}{p-1}} \tau_p(\mathbf{x})^{\binom{p^n}{p}} \cdots .$$

Since $\tau_i(\mathbf{x})$ is a product of commutators in $x_1, \cdots, x_m$ of weight at least i, $\tau_i(\mathbf{x}) = 1$ if $i \geq p$. Now as we noted in 0.1, if $g \in G$ then $gp(g, G')$ is of class at most $c - 1$. In particular

$$gp(\tau_1(\mathbf{x}), \tau_2(\mathbf{x}), \cdots, \tau_{p-1}(\mathbf{x}))$$

is of class at most $c - 1$. Since each binomial coefficient $\binom{p^n}{i}$ is divisible by p^n when $1 \leq i \leq p - 1$, the induction assumption enables us to conclude that

$$\tau_1(\mathbf{x})^{p^n} \tau_2(\mathbf{x})^{\binom{p^n}{2}} \cdots \tau_{p-1}(\mathbf{x})^{\binom{p^n}{p-1}}$$

is a p^nth power. This completes the proof of Corollary 2.31.

We need a similar result for primes which may not exceed the class.

Corollary 2.32 (N. Blackburn [14]). *For each prime p and each positive integer c there exists an integer $f(p, c)$ such that if G is a nilpotent group of class at most c, then every product of p^nth powers (for any $n \geq f(p, c)$) of elements of G is a $p^{n-f(p,c)}$th power.*

Proof. Once the function $f(p, c)$ has been defined correctly, the argument is similar to the one given above. The details of the proof may be found in [14].

2.5. We shall make use of Corollary 2.31 in order to prove a useful result which is related to the aims set forth at the beginning of 2.4. (Much of the discussion in 2.5, 2.6 and 2.7 is contained in the papers [8] and [9], although a fair amount of it is new.)

We recall some notation which we introduced in 0.5. Given a prime p, a subgroup H of a group G is *p-isolated in* G if whenever $g^p \in H$, $g \in G$, then $g \in H$. If H is p-isolated for every prime p, we call H an *isolated* subgroup of G.

Here we prove the

Lemma 2.6. *Let G be a finitely generated nilpotent group and let H be a subgroup of G. Then H is p-isolated in G for all but finitely many primes p.*

Proof. If G is abelian, then G/H is the direct product of a finite group of order n, say, and a free abelian group. Thus if p is any prime bigger than n then H is p-isolated.

Thus we may assume that the class c of G is at least 2. Inductively HZ/Z is p-isolated in G/Z for all but a finite set Ω of primes, where $Z = \zeta G$. So if $p \notin \Omega$, if $g \in G$ and $g^p \in HZ$ then $g \in HZ$. Now H is normal in HZ and

$$HZ/H \cong Z/Z \cap H.$$

It follows as in the abelian case that if p is any prime which exceeds the order of the torsion subgroup of $Z/Z \cap H$ then H is p-isolated in HZ.

Let Λ be the set of all primes which are not only outside Ω but which also exceed the order of the torsion subgroup of $Z/Z \cap H$. Since a p-isolated subgroup of a p-isolated subgroup is p-isolated in the whole group, H is p-isolated in G for every p in Λ. This completes the proof.

If G is a group and m is an integer, we denote by G^m the subgroup of G generated by the mth powers of its elements. Putting Lemma 2.6 together with Corollary 2.31 we have the

Proposition 2.2. *Let G be a finitely generated nilpotent group and let H be a subgroup of G. Then for all but a finite number of primes p and for every positive integer n*

$$H \cap G^{p^n} = \{h^{p^n} | h \in H\}.$$

We proceed now to a proof of the uniqueness of the embedding arrived at in Lemma 2.3. We prove first

Lemma 2.7. *Let A and B be finitely generated torsion-free nilpotent groups, let H and K be torsion-free nilpotent supergroups of A and B respectively, and let $x \in H$ and $y \in K$ be such that*

$$H = gp(A, x), \; K = gp(B, y).$$

Let n be a positive integer such that

$$x^n = a \in A \text{ and } y^n = b \in B.$$

If ϕ is a homomorphism of A onto B which maps a onto b then ϕ can be extended to a homomorphism of H onto K in such a way that x is mapped onto y.

Proof. Suppose A is generated by $a_1, \cdots, a_k$; then

$$H = gp(a_1, \cdots, a_k, x), \quad K = gp(a_1\phi, \cdots, a_k\phi, y).$$

We shall show that every relation between $a_1, \cdots, a_k, x$ is also a relation between $a_1\phi, \cdots, a_k\phi, y$. Thus assume

$$w(a_1\phi, \cdots, a_k\phi, y) \neq 1;$$

we have to prove

$$w(a_1, \cdots, a_k, x) \neq 1. \tag{1}$$

To do so we choose by Proposition 2.2 a prime p which does not divide n such that

$$A^{p^i} = H^{p^i} \cap A, \tag{2}$$

for $i = 1, 2, \cdots$. Since $K \in \mathbf{R}\mathfrak{F}_p$ we can find an integer i for which

$$w(a_1\phi, \cdots, a_k\phi, y) \notin K^{p^i}. \tag{3}$$

Now p^i and n are coprime. So by Euclid's algorithm we can find u and v such that

$$up^i + vn = 1. \tag{4}$$

Therefore, working modulo K^{p^i}, we have

$$y = y^{up^i + vn} = (y^n)^v = b^v.$$

So, by (3), $w(a_1\phi, \cdots, a_k\phi, b^v) \notin K^{p^i}$, and in particular $w(a_1\phi, \cdots, a_k\phi, b^v) \notin B^{p^i}$. But ϕ maps A homomorphically onto B. Consequently $w(a_1, \cdots, a_k, a^v) \notin A^{p^i}$, which, by (2), implies that $w(a_1, \cdots, a_k, a^v) \notin H^{p^i}$. However, working modulo H^{p^i}, we have $a^v = x^{nv} = x^{1-up^i} = x$. This means $w(a_1, \cdots, a_k, x) \notin H^{p^i}$, and so in particular $w(a_1, \cdots, a_k, x) \neq 1$. The lemma now follows immediately from the well-known theorem of von Dyck.

An immediate, important consequence of Lemma 2.7 is the following

Corollary 2.71. *Let A, B, H and K be as above. If ϕ is an isomorphism then so is its extension to H.*

It follows from Corollary 2.71 that there is essentially only one way to adjoin an nth root to an element in a finitely generated torsion-free nilpotent group if the resultant group is also torsion-free nilpotent. This remark enables us to prove the following theorem of A. I. Mal'cev [63].

Theorem 2.4. *Let A be a torsion-free nilpotent group. Then A can be embedded in a torsion-free nilpotent $\mathfrak{D}$-group.*

Proof. It is enough to show that if p is any prime and a any element of A then we can find a torsion-free nilpotent group B containing A in which the element a has a pth root.

To do so we choose a system of generators $\{a_i | i \in I\}$ for A. Now a is a word in the generators a_i: $a = w(a_1, \cdots, a_n)$.

It follows from Corollary 2.71 that for each subgroup of A generated by a finite subset of the given generators of A which contains $a_1, \cdots, a_n$, there is a unique way of adjoining a pth root, always labelled x, to a in a torsion-free nilpotent group. If one describes such groups in terms of generators and defining relations then the resultant system of generators

$$\{a_i | i \in I\} \cup \{x\}$$

together with the acquired system of relations clearly defines a torsion-free nilpotent group containing A in which a has a pth root x—since any relation between these elements arises from a relation in some finitely generated subgroup. This completes the proof.

Let us term a minimal torsion-free nilpotent $\mathfrak{D}$-group $m(G)$ containing a given torsion-free nilpotent group G a Mal'cev completion of G. An immediate consequence of our approach to the existence of an $m(G)$ (see in particular Lemma 2.7) is the following generalization of a theorem (viz. Corollary 2.42 below) of Mal'cev [63].

Corollary 2.41. *Let G, H be torsion-free nilpotent groups and let ϕ be a homomorphism of G into H. If $m(G)$ and $m(H)$ are any Mal'cev completions of G and H, th then ϕ can be extended uniquely to a homomorphism $m(\phi)$ of $m(G)$ onto $m(H)$.*

Corollary 2.41 yields the further

Corollary 2.42. *Let G be a torsion-free nilpotent group. If $m(G)$ and $m'(G)$ are Mal'cev completions of G then they are isomorphic.*

2.6. Corollary 2.42 completes one aspect of the applications of residual properties to nilpotent groups. In 2.7 we shall give another application which depends on the following

Theorem 2.5. *Let G be a finitely generated torsion-free nilpotent group and let H be an isolated subgroup of G. Then for any given prime* p,

$$\bigcap_{i=1}^{\infty} G^{p^i}H = H.$$

It is worth noting that Theorem 2.5 is a generalization of Gruenberg's theorem (the case $H = 1$).

For the proof of Theorem 2.5, we need two lemmas, the first of which is due to A. I. Mal'cev.

Lemma 2.8. *Let G be a finitely generated nilpotent group and let H be a subgroup of G. If some positive power of each element of a set of generators of G lies in H, then H is of finite index in G and a positive power of every element of G lies in H.*

Proof. We proceed by induction on the class c of G. If $c = 1$ the result is clear. Suppose $c > 1$; then by induction $[G: H\lambda_c G] < \infty$ and so a positive power of every element of G lies in $H\lambda_c G$.

Let $\{g_1, g_2, \cdots, g_s\}$ be a set of generators for G, and suppose $g_i^{m_i} \in H$, where $m_i > 0$, $1 \le i \le s$. Let $\{c_1, c_2, \cdots, c_t\}$ be a set of generators for $\lambda_{c-1}G$ and suppose $c_j^{n_j} \in H\lambda_c G$, where $n_j > 0$, $1 \le j \le t$. Since $\lambda_c G \le \zeta G$, it follows at once from the commutator formulae (Theorem 0.2 (i)) that

$$\lambda_c G = gp([c_j, g_i] | 1 \le i \le s, 1 \le j \le t).$$

Similarly

$$[c_j, g_i]^{n_j m_i} = [c_j^{n_j}, g_i^{m_i}] \in H \ (1 \le i \le s, 1 \le j \le t).$$

Thus the abelian group $H\lambda_c G/H$ is finite. This completes the proof.

If G is any group and H is a subgroup of G, the *isolator* $i(H)$ *of* H *in* G is the intersection of all the isolated subgroups of G containing H. Since the intersection of isolated subgroups is clearly isolated, $i(H)$ is the unique minimal isolated subgroup of G containing H.

If G is a nilpotent group and H is a subgroup of G, it follows from the preceding lemma that

$$J = \{g \in G | \text{ some positive power of } g \text{ lies in } H\}$$

is a subgroup of G. Therefore $J = i(H)$.

Lemma 2.9. *Let G be a finitely generated torsion-free nilpotent group, let H be an isolated subgroup of G and let Z be a subgroup of the center of G. If the isolator of HZ in G is G, then H is normal in G.*

Proof. We proceed by induction on the torsion-free rank r of G. If $r = 1$, G is abelian and H is certainly normal in G.

Suppose $r > 1$. By Lemma 2.1, G is a $\mathfrak{U}$-group, so that, by Theorem 0.4, all centralizers are isolated in G. In particular, since

$$\zeta H = H \cap (\text{centralizer of } H \text{ in } G),$$

ζH is isolated in G.

By our remarks preceding this lemma, if $g \in G$ then some power g^m $(m > 0)$ of g lies in HZ; so g^m centralizes ζH, and therefore so does g. This means $\zeta H \leq \zeta G$.

Now $G/\zeta H$ is torsion-free because ζH is isolated in G, so by Theorem 0.1, $G/\zeta H$ has torsion-free rank less than r. The induction assumption then gives that $H/\zeta H$ is normal in $G/\zeta H$, so that H is normal in G. This completes the proof of the lemma.

We are now in a position to proceed with the

Proof of Theorem 2.5. Let G be a finitely generated torsion-free nilpotent group, let H be an isolated subgroup of G and let p be any prime. We proceed by induction on the class c of G.

If $c = 1$, G is a free abelian group of finite rank. Since H is isolated in G, G/H is torsion-free and so is also free abelian; therefore $G = H \times K$, for suitable choice of $K \leq G$. It follows that

$$G^{p^i}H = H \times K^{p^i}, \; i = 1, 2, \cdots .$$

Since K is free abelian, $\bigcap_{i=1}^{\infty} K^{p^i} = 1$, so

$$\bigcap_{i=1}^{\infty} G^{p^i}H = H.$$

Suppose now that $c > 1$. We put

$$Z = \zeta G, \; I = i(HZ) \text{ and } L = \bigcap_{i=1}^{\infty} G^{p^i}H.$$

By Corollary 2.11, G/Z is torsion-free nilpotent, and I/Z is an isolated subgroup of G/Z. Inductively

$$\bigcap_{i=1}^{\infty} G^{p^i}I = I,$$

so that certainly $\bigcap_{i=1}^{\infty} G^{p^i}H \leq I$. Thus $H \leq L \leq I$. Applying Lemma 2.9 to the finitely generated torsion-free nilpotent group I with isolated subgroup H, we deduce that H is normal in I. Also I/H is torsion-free, since H is isolated in I.

We claim that in fact $L/H = 1$ i.e. that $L = H$. For suppose $lH \in L/H$. Then given any integer $i > 0$ we can find elements $g_1, \cdots, g_t \in G$, $h \in H$ such that $l = g_1^{p^i} \cdots g_t^{p^i} h$. If i is sufficiently large then, by Corollary 2.32 $g_1^{p^i} \cdots g_t^{p^i}$ can be written as p^j-power, say g^{p^j}, where j tends to infinity with i. Thus $g^{p^j} = lh^{-1} \in I$. Since I is isolated, $g \in I$. Thus we have

$$l = g^{p^j} h \ (h \in H).$$

This means l is modulo H, a p^nth power (in I) for every $n \geq 1$, i.e. lH has a p^nth root in I/H for every n. So if $lH \neq H$ we can find a properly increasing infinite series of subgroups in the finitely generated torsion-free nilpotent group I/H. But I/H satisfies the maximal condition. Hence $lH = H$ is the only possibility and so $L = H$ as desired.

2.7. We are now able to give a new proof of the following theorem of V. M. Glus̆kov [52], p. 249) by a simple application of Theorem 2.5.

Theorem 2.6. *The normalizer N of an isolated subgroup H of a finitely generated torsion-free nilpotent group G is an isolated subgroup.*

Proof. Suppose N is not isolated. Then there exists a prime p and an element x in G such that $x^p \in N$ but $x \notin N$. Therefore there must be an element $h \in H$ such that $x^{-1}hx \notin H$. Hence, by Theorem 2.5, given any prime q there exists an integer $i \geq 1$ such that

$$x^{-1}hx \notin G^{q^i} H. \tag{1}$$

Let us choose $q \neq p$. Then modulo G^{q^i}, x is actually a power of x^p. Hence modulo G^{q^i}, x normalizes H, which implies $x^{-1}hx \in G^{q^i}H$. This contradicts (1) and so the proof of Glus̆kov's theorem is complete.

2.8. We come finally to a rather different application of the results we have obtained so far. This application is based on the well-known

Proposition 2.3. *A finite extension of a nilpotent $\mathfrak{D}$-group splits.*

Proof. Suppose we have a short exact sequence

$$1 \longrightarrow N \longrightarrow G \longrightarrow F \longrightarrow 1 \tag{1}$$

with N a nilpotent $\mathfrak{D}$-group and F finite. Observe that it suffices to prove that there is a subgroup E of G such that $|E| = |F|$.

This is done most easily by induction on the class of N. If N is abelian then the sequence (1) splits because N is a torsion-free abelian group in which extraction of nth roots is possible for every $n > 0$ and F is finite ([58], Corollary 5.4). Thus we can find a subgroup of order $|F|$.

Suppose N is not abelian and let Z be the center of N. By Theorem 0.4 and Corollary 0.41, Z is itself a $\mathfrak{D}$-group and N/Z is a nilpotent $\mathfrak{D}$-group of class less than the class of N. Inductively we can find a subgroup E_1/Z of G/Z of order $|G/N|$. But now E_1 is an extension of the abelian $\mathfrak{D}$-group Z by the finite group

E_1/Z. Thus E_1 has a subgroup E, say, of order $|E_1/Z| = |F|$. This completes the proof of the proposition.

Proposition 2.3 is the key fact in our proof of the following theorem of R. Baer [5].

Theorem 2.7. *Let G be a group such that $G/v_k G$ is finite. Then $\lambda_{k+1} G$ is finite.*

Proof. We proceed by induction on k, the result being immediate for $k = 0$.

Let k be a positive integer and let G be a group such that $G/v_k G$ is finite. Applying the induction assumption to the group $G/\zeta G$ it follows that $\lambda_k G/\lambda_k G \cap \zeta G$ is finite. Combining this fact with the finiteness of $G/v_k G$ and the identity $[v_k G, \lambda_k G] = 1$, it is easy to see that $\lambda_{k+1} G$ is finitely generated. Thus we may assume G is finitely generated.

Put $U = v_k G$; then U is finitely generated since it is of finite index in a finitely generated group. So U is a finitely generated nilpotent group. Therefore U is residually finite. Hence we can find a normal torsion-free nilpotent subgroup N of U of finite index. Now U is of finite index in G. So we can also arrange N to be normal in G.

Let us now choose a system $T = \{t_1, t_2, \cdots, t_l\}$ of representatives of the cosets of N in G. We denote the representative of the coset gN $(g \in G)$ by $\tilde{g}$; thus $\tilde{g} = \tilde{h}$ if and only if g and h lie in the same coset. Every element g in G can be written uniquely in the form

$$g = tn \ (t \in T,\ n \in N).$$

Associated with this system of coset representatives T, there is a function $f\colon T \times T \dashrightarrow N$ given by

$$t_i t_j = \widetilde{t_i t_j} f(t_i, t_j) \ (t_i, t_j \in T).$$

Finally observe that each representative t_i gives rise to an automorphism γ_i of N, via conjugation,

$$\gamma_i\colon n \longrightarrow t_i^{-1} n t_i \ (n \in N).$$

These definitions allow us to express the composition in G as follows:

$$t_i m \cdot t_j n = \widetilde{t_i t_j} f(t_i, t_j)(m\gamma_j)n \ (t_i, t_j \in T,\ m,\ n \in N).$$

Now let $\bar{N}$ be a Mal'cev completion of N. Every automorphism α of N extends uniquely to an automorphism $\bar{\alpha}$ of $\bar{N}$. Let $\bar{G}$ be the set of all pairs tn, $t \in T$, $n \in \bar{N}$, and define

$$t_i m \cdot t_j n = \widetilde{t_i t_j} f(t_i, t_j)(m\bar{\gamma}_j)n \ \ (t_i, t_j \in T,\ m,\ n \in \bar{N}).$$

Then $\bar{G}$ is a group and both G and $\bar{N}$ may be regarded as subgroups of $\bar{G}$ with $\bar{N}$

normal in $\bar{G}$. Further $G \cap \bar{N} = N$, $G\bar{N} = \bar{G}$ and the set T may be regarded as a system of coset representatives for $\bar{G}$ modulo $\bar{N}$. It follows that

$$\bar{G}/\bar{N} \text{ is finite of order } l.$$

Thus, by Proposition 2.3, $\bar{G}$ splits over $\bar{N}$. We can therefore find elements $n_1, n_2, \ldots, n_l \in \bar{N}$ so that

$$\{t_1 n_1, t_2 n_2, \ldots, t_l n_l\} = E$$

is a subgroup of $\bar{G}$ of order l.

Recalling that $N \leq v_k G$, we claim that $\bar{N} \leq v_k \bar{G}$. To see this, we note first that, as an immediate consequence of the remarks following Lemma 2.8, every element $\bar{n}$ of the Mal'cev completion $\bar{N}$ of N has the property that some positive power of $\bar{n}$ lies in N. Consider the series

$$N = N_1 \geq N_2 \geq \cdots \geq N_{k+1} = 1,$$

where $N_{i+1} = [N_i, G]$. From this we construct a series of subgroups of $\bar{N}$:

$$\bar{N} = \bar{N}_1 \geq \bar{N}_2 \geq \cdots \geq \bar{N}_{k+1} = 1,$$

where $\bar{N}_i$ is the isolator of N_i in $\bar{N}$ (i.e. the $\mathfrak{D}$-closure of N_i). To establish $\bar{N} \leq v_k \bar{G}$, it suffices to prove

$$[\bar{N}_i, \bar{G}] \leq \bar{N}_{i+1}, \quad i = 1, 2, \ldots, k. \tag{1}$$

In order to prove (1) notice that N_k is central in N and therefore $\bar{N}_k$ is central in $\bar{N}$. Let $a \in \bar{N}_k$, $g \in \bar{G}$. Then $a^m \in N_k$ for a suitable choice of $m > 0$. Hence

$$(g^{-1}ag)^m = g^{-1}a^m g = a^m. \tag{2}$$

Of course $g^{-1}ag \in \bar{N}$ and $\bar{N}$ is a U-group. So it follows from (2) that $g^{-1}ag = a$. This means $[\bar{G}, \bar{N}_k] = 1$ as desired. The rest of (1) follows by induction on considering $\bar{G}/\bar{N}_k$.

Thus we have $\bar{G} = E\bar{N}$, where $\bar{N} \leq v_k \bar{G}$. It follows that E actually centralizes $\bar{N}$. For suppose that $[\bar{N}, E] \neq 1$. Then there is an integer $j \geq 1$ such that $M_j \neq 1$, $M_{j+1} = 1$, where

$$M_i = [\bar{N}, \underbrace{E, \ldots, E}_{i}], \quad i = 0, 1, 2, \ldots.$$

So E centralizes M_j but does not centralize M_{j-1}. Let $n \in M_{j-1}$ be such that $[E, n] \neq 1$. Then the mapping

$$e \longrightarrow [e, n], \; (e \in E),$$

is a homomorphism of E into $\bar{N}$. As $\bar{N}$ is torsion-free, the image of E under this homormorphism must be the identity, which contradicts $[E, n] \neq 1$.

Therefore in fact $\bar{G} = E \times \bar{N}$. Hence, remembering that $\lambda_{k+1}\bar{N} = 1$, we have $\lambda_{k+1}\bar{G} = \lambda_{k+1}E$. Thus $\lambda_{k+1}\bar{G}$ is finite and therefore so is $\lambda_{k+1}G$. This completes the proof of Theorem 2.7.

For some allied results in this direction see D. J. S. Robinson [80].

There is an easy converse of Theorem 2.7, namely the following

Theorem 2.8. *Let G be a finitely generated group such that $\lambda_{k+1}G$ is finite. Then G/v_kG is finite.*

Actually Theorem 2.8 is due to P. Hall (according to a written communication from K. W. Gruenberg). If we remove the condition that G be finitely generated, we can conclude only that $G/v_{2k}G$ is finite.

Theorem 2.8′. *Let G be a group such that $\lambda_{k+1}G$ is finite. Then $G/v_{2k}G$ is finite.*

Theorem 2.8′ may be found in the paper by P. Hall [33], where examples are constructed to show that the number $2k$ is best possible.

Here we shall prove only Theorem 2.8. To this end let C be the centralizer of $\lambda_{k+1}G$ in G. Since $\lambda_{k+1}G$ is a finite normal subgroup of G, C is normal and of finite index in G. Clearly $\lambda_{k+1}C$ is central in C, so C is nilpotent. Moreover C is finitely generated because G is. Thus we can find a torsion-free normal subgroup N of C of finite index, and as C is of finite index in G, we may assume that N is also normal in G. Now

$$[N, \underbrace{G, \dots, G}_{k}] \leq \lambda_{k+1}G \cap N.$$

Since $\lambda_{k+1}G$ is finite and N is torsion-free,

$$[N, \underbrace{G, \dots, G}_{k}] = 1.$$

In other words

$$N \leq v_k G.$$

Since N is of finite index in G, we find G/v_kG is finite as desired.

Chapter 3. Lie and associative ring techniques and the commutator calculus

In this chapter we introduce the commutator calculus keeping the catalytic lie and associative rings in the foreground. Using the notion of basic sequence introduced by P. Hall (see e.g. [32]) we shall prove a theorem of K. W. Gruenberg [24]. We begin with a brief discussion of freeness.

3.1. Suppose Ω is a (possibly empty) set such that to each $\omega \in \Omega$ there is assigned a nonnegative integer $\nu(\omega)$, its *arity*. The elements of Ω are termed *operators*. An Ω-group is a group G in which there is defined for each $\omega \in \Omega$ of arity n an n-ary operation (again denoted ω) in G, that is, a mapping of the set-theoretical product of n copies of G into G. Thus an Ω-group is simply a group with operators (where here the operators need not, of course, be endomorphisms). This point of view allows us to speak of Ω-subgroups, normal Ω-subgroups, Ω-factor groups and so on. If G and H are Ω-groups, a *homomorphism* ϕ of G into H is a group-theoretical homomorphism of G into H with the additional property that for each $\omega \in \Omega$

$$((a_1, a_2, \cdots, a_n)\omega)\phi = (a_1\phi, a_2\phi, \cdots, a_n\phi)\omega \quad (\nu(\omega) = n)$$

for every choice of elements $a_1, a_2, \cdots, a_n \in G$. A nonempty class $\underline{\underline{V}}$ of Ω-groups is termed a *variety* if it is closed under the formation of Ω-subgroups, Ω-factor groups and cartesian products.

Suppose next that $\underline{\underline{V}}$ is any class of Ω-groups. We term an Ω-group G a $\underline{\underline{V}}$-group if $G \in \underline{\underline{V}}$. A $\underline{\underline{V}}$-group F is *free* (in $\underline{\underline{V}}$) if there is a subset X of F such that for every $\underline{\underline{V}}$-group G and every mapping θ of X into G there is a unique homomorphism (of Ω-groups) ϕ of F into G which agrees with θ on X; in such a situation we say X *freely generates* F. If $\underline{\underline{V}}$ is a variety of Ω-groups with the property that $\underline{\underline{V}}$ does not consist of groups of order one then, according to a theorem of G. Birkhoff, for every cardinal number m, there exists a free $\underline{\underline{V}}$-group freely generated by a set of m elements.

Now suppose $\Omega = \emptyset$, the empty set. Then an Ω-group is simply a group. Let $\underline{\underline{V}}$ be the variety of all groups. In this variety there then exist, of course, free groups. Suppose F is such a free group freely generated by X. Then every element $f \in F$ $(f \neq 1)$ can be written uniquely as a *reduced X-product*, i.e., a product $a_1 a_2 \cdots a_n$ where $a_i \in X \cup X^{-1}$ and no pair $a_i a_{i+1}$ is of the form xx^{-1} or $x^{-1}x$ $(x \in X)$. $F/\lambda_2 F$ is a direct product of $|X|$ infinite cyclic groups, i.e., $F/\lambda_2 F$ is free abelian of rank $|X|$; $|X|$ is an invariant of F which is often termed the *rank of* F.

Similarly if we take Ω to consist of a single operation ω of arity 2 and denote by $\underline{\underline{V}}$ the class of all Ω-groups in which the group multiplication is commutative and both distributive laws and the associative law (of ω-multiplication) hold, a $\underline{\underline{V}}$-group turns out to be a ring. Thus the class of all rings constitutes a variety. Suppose R is a free ring freely generated by X. Then two X-monomials $x_1x_2 \cdots x_n$ and $y_1y_2 \cdots y_m$ $(x_i, y_j \in X)$ are equal only if $m = n$ and $x_1 = y_1$, $x_2 = y_2, \ldots, x_n = y_n$; in addition every element r $(r \neq 0)$ can be written uniquely as a linear sum

$$r = \sum u_{i_1,i_2,\ldots,i_n} x_{i_1} x_{i_2} \cdots x_{i_n}$$

of integer multiples of distinct monomials (i.e. $u_{i_1,i_2,\ldots,i_n} \in Z$, the ring of integers, and $u_{i_1,i_2,\ldots,i_n} \neq 0$). We allow $n = 0$, i.e., the empty monomial is included in R which means that R is a ring with 1.

The analogous comments hold also for associative algebras over a field.

We recall that a ring R is *graded* if R is the direct sum of its additive subgroups $R_0, R_1, \cdots$ where $R_iR_j \subseteq R_{i+j}$:

$$R = \sum_{i=0}^{\infty} R_i.$$

The elements of R_i are termed *homogeneous* of degree i. If $r \in R$ then $r = \sum_{i=0}^{\infty} r_i$ where almost all the $r_i = 0$. We define the completion $\bar{R}$ of a graded ring $R = \sum_{i=0}^{\infty} R_i$ to consist of the (cartesian) product of the abelian groups R_i:

$$\bar{R} = \prod_{i=0}^{\infty} R_i.$$

We denote an element of $\bar{R}$ by $\sum_{i=0}^{\infty} r_i$ if its component in R_i is r_i for $i = 0, 1, \cdots$. The multiplication in $\bar{R}$ is as usual

$$\sum_{i=0}^{\infty} r_i \cdot \sum_{i=0}^{\infty} s_i = \sum_{i=0}^{\infty} \left[\sum_{k+l=i} r_k s_l \right].$$

Similarly for algebras over a field (where vector subspaces take the place of additive subgroups).

Let I be a nonempty index set and let R be the free ring on $Y = \{y_i | i \in I\}$. $R = \sum_{i=0}^{\infty} R_i$ where R_i is the free abelian group on the products $y_{j_1} \cdots y_{j_i}$ $(j_k \in I)$. Clearly R is a graded ring. We term $\bar{R}$ the *Magnus power series* ring on Y. By our earlier remark the empty monomial is included and so R_0 may be identified with Z, the ring of integers. We write $\bar{R} = Z[[Y]]$.

For each $r \in Z[[Y]]$, its *order* $O(r)$ is defined by

$$O(r) = \begin{cases} \min\{i \mid r_i \neq 0\} & \text{if } r \neq 0, \\ \infty & \text{if } r = 0. \end{cases}$$

For each $i = 0, 1, 2, \cdots$ we define an ideal A_i of $\bar{R}$ by

$$A_i = \{r \in Z[[Y]] \mid O(r) \geq i\}.$$

It is clear that

$$\bigcap_{i=0}^{\infty} A_i = 0.$$

Lemma 3.1. *The group U of units of $Z[[Y]]$ is residually nilpotent. U contains a subgroup V of index two which is residually torsion-free nilpotent.*

Proof. It is easy to see that $U = \{r \in Z[[Y]] \mid r_0 = \pm 1\}$. We define $V = \{r \in Z[[Y]] \mid r_0 = +1\}$. It is clear that V is of index two in U. Put

$$U(i) = \{\pm 1 + \alpha \in U \mid O(\alpha) \geq i\}, \quad i = 1, 2, \cdots,$$

and define $V(i)$ similarly. It is clear that each $U(i)$ ($V(i)$) is a normal subgroup of U (V). We shall prove by induction on i that $\lambda_i U \leq U(i)$ ($\lambda_i V \leq V(i)$).

Assuming $\lambda_{i-1} U \leq U(i-1)$, let $\pm 1 + \alpha \in U$, $\pm 1 + \beta \in \lambda_{i-1} U$ so that $O(\beta) \geq i - 1$. Then

$$\begin{aligned}[\pm 1 + \alpha, \pm 1 + \beta] &= (\pm 1 + \alpha)^{-1}(\pm 1 + \beta)^{-1}(\pm 1 + \alpha)(\pm 1 + \beta) \\ &= (\pm 1 + \alpha)^{-1}(\pm 1 + \beta)^{-1}\{(\pm 1 + \beta)(\pm 1 + \alpha) \\ &\quad + (\pm 1 + \alpha)(\pm 1 + \beta) - (\pm 1 + \beta)(\pm 1 + \alpha)\} \\ &= 1 + (\pm 1 + \alpha)^{-1}(\pm 1 + \beta)^{-1}(\alpha\beta - \beta\alpha).\end{aligned}$$

Therefore

$$O(1 - [\pm 1 + \alpha, \pm 1 + \beta]) \geq O(\alpha) + O(\beta) \geq 1 + (i - 1) = i,$$

and $\lambda_i U \leq U(i)$ is established by induction (and similarly for V).

Now $\bigcap_{i=0}^{\infty} A_i = 0$; so it is clear that each nontrivial element of U lies outside $U(i)$ for some i. This means U is residually nilpotent. To complete the proof of Lemma 3.1 it remains only to remark that $V/V(i)$ is torsion-free.

3.2. **Some important properties of free groups.** The relevance of the above discussion to free groups arises from the following

Theorem 3.1 (**W. Magnus** [59]). *Every free group has a faithful representation in a*

suitably chosen Magnus power series ring.

Proof. We claim that the subgroup $gp(1 + y_i | i \in I)$ of the group of units U of $Z[[Y]]$ is in fact a free group freely generated by these elements. To verify this it suffices to prove that an arbitrary "contracted" reduced product

$$p = (1 + y_{i_1})^{n_1}(1 + y_{i_2})^{n_2} \dots (1 + y_{i_k})^{n_k},$$

where $k \geq 1$, $n_j \neq 0$, $1 \leq j \leq k$, and $i_j \neq i_{j+1}$, $1 \leq j < k$, is never equal to 1. But it is not difficult to see that

$$p = (1 + n_1 y_{i_1} + \dots)(1 + n_2 y_{i_2} + \dots) \dots (1 + n_k y_{i_k} + \dots)$$

$$= 1 + \dots + n_1 n_2 \cdots n_k y_{i_1} y_{i_2} \cdots y_{i_k} + \cdots .$$

It follows by inspection that there is no other term that arises on computing p which matches $n_1 n_2 \cdots n_k y_{i_1} y_{i_2} \cdots y_{i_k}$. So $p \neq 1$ as required.

From this representation and Lemma 3.1 it follows that we have proved another theorem of W. Magnus [59].

Theorem 3.2. *Free groups are residually torsion-free nilpotent.*

This theorem together with Gruenberg's Theorem (Theorem 2.2) implies the following result of K. Iwasawa.

Corollary 3.21. *Let F be a free group and let p be any prime. Then $F \in \mathbf{R}\mathfrak{F}_p$.*

Let $\mathfrak{N}$ denote the class of all nilpotent groups. Then we have the following

Proposition 3.1. *A finitely generated $\mathbf{R}\mathfrak{N}$ group is hopfian.*

Proof. Let G be finitely generated and residually nilpotent and let η be an epimorphism of G to G. Suppose $g \neq 1$ belongs to the kernel of η. Then for some n, $g \notin \lambda_n G$. But η induces an epimorphism $\hat{\eta}$: $G/\lambda_n G \to G/\lambda_n G$ which is an isomorphism since finitely generated nilpotent groups are hopfian (Corollary 0.11). This contradicts the fact that $g\lambda_n G$ is a nontrivial element in $\ker \hat{\eta}$. This completes the proof of the proposition.

There are two consequences I would like to deduce. The first is due to J. Nielsen [75], while the second is due to W. Magnus (see [61], p. 346).

Theorem 3.3. *Finitely generated free groups are hopfian.*

We say that a group G *has the same lower central sequence as a free group F* if $G/\lambda_n G \cong F/\lambda_n F$ for every n.

Theorem 3.4. *Let m be a positive integer. An m-generator group G with the same lower central sequence as a free group F of rank m is free.*

Proof. Let ϵ be an epimorphism of F onto G. If $1 \neq f \in F$ then $f \notin \gamma_n F$ for some n. Let $\hat{\epsilon}$ be the epimorphism of $F/\gamma_n F$ onto $G/\gamma G$ induced by ϵ. Since

$F/\gamma_n F \cong G/\gamma_n G$, $\hat{\epsilon}$ is an isomorphism, so that $f\epsilon \notin \gamma_n G$. Therefore $f \notin \ker \epsilon$ and ϵ is an isomorphism as required.

Magnus' theorem invites the

Definition. A group G is termed *parafree* if

(a) $G \in \mathbf{R}\mathfrak{N}$,

(b) G has the same lower central sequence as a free group.

In particular free groups are parafree. Theorem 3.4 suggests that nonfree parafree groups may be hard to find. In fact there are many of them. It suffices here here to exhibit one of them, viz.

$$G = |a, b, c; a^2 b^3 c^5|.$$

The notation we are using is the following: If X freely generates a free group F and R is a subset of F then $|X; R|$ denotes the factor group F/N, where $N = gp_F(R)$ is, as usual, the least normal subgroup of F containing R. Thus $G = F/gp_F(a^2b^3c^5)$ where F is free on a, b, c. This group G cannot be generated by two elements; it is residually nilpotent and has the same lower central sequence as a free group of rank two. Thus G is parafree, but not free.

It turns out parafree groups share many of the properties of free groups. Thus, for example, the frattini subgroup of a parafree group is trivial. Thinking along slightly different lines one finds that the two-generator subgroups of parafree groups are again free.

3.3. Basic commutators. We will describe here the theory of basic commutators on a *finite* set X. Thus throughout this section, let X denote a nonempty finite set:

$$X = \{x_1, x_2, \cdots, x_q\}.$$

Let G be the *free groupoid* freely generated by the set X. Thus G is a nonempty set equipped with single binary operation; we write gh $(g, h \in G)$ for the product of g and h relative to this binary operation. The elements of G are simply the bracketed products of the elements of X; two such products are equal if and only if they are identical. Every element g in G is uniquely a product of elements in X; the number of factors is termed the *length* of g and denoted $|g|$. For example if $x, y \in X$,

$$|x| = 1, \ |xy| = 2, \ |x(xy)| = 3,$$

and so on. Thus every element g in G of length greater than 1 is uniquely the product $g = g'g''$ of two elements $g', g'' \in G$ of shorter length.

A sequence $b_1, b_2, \cdots$ of elements of the free groupoid G on X is termed a *basic sequence in* X if

(i) every element of X occurs in the sequence;

(ii) if $|b_i| < |b_j|$ then $i < j$;

(iii) if $u = vw(v, w \in G)$ is an element of G of length at least two, then u occurs in the sequence (1) if and only if

(a) $v = b_i$, $w = b_j$ and $j < i$, and

(b) either $|b_i| = 1$ or $b_i = b_k b_l$ and $l \leq j$.

The existence of a basic sequence in X is taken care of by the following

Proposition 3.1. *Let $X = \{x_1, \cdots, x_q\}$ and let G be the free groupoid on X. Then there exists a basic sequence in X.*

Proof. Define a total order in X, e.g. the one consistent with the labelling of the elements of X. Now let L_n denote the set of elements of G of length n for $n = 1, 2, \cdots$. Since X is finite, so is L_n $(n = 1, 2, \cdots)$. Suppose $n > 1$ and that $B_1 \subseteq L_1, \cdots, B_{n-1} \subseteq L_{n-1}$ have been defined and totally ordered so that (i), (ii) and (iii) hold for elements of length at most $n - 1$ in $\bigcup_{i=1}^{n-1} B_i$. The set B_n is then unambiguously defined by (iii), it is totally ordered arbitrarily and connected to the B_i $(i < n)$ by defining $u < v$ if $u \in B_i$ $(i < n)$ and $v \in B_n$. It is clear that $B = \bigcup_{n=1}^{\infty} B_n$, subject to the imposed ordering, is a basic sequence.

More generally suppose that Γ is a groupoid generated by $\Xi = \{\xi_1, \cdots, \xi_q\}$. A sequence $\beta_1, \beta_2, \cdots$ of elements of Γ is termed a basic sequence in Ξ if there is a basic sequence $b_1, b_2, \cdots$ in X such that the homomorphism of G onto Γ defined by $x_i \dashrightarrow \xi_i$ $(i = 1, \cdots, q)$ maps b_i onto β_i for every i.

Proposition 3.1 provides us with one way of obtaining a basic sequence in X. We shall need a somewhat different means for producing a basic sequence. To this end let us revert to the groupoid Γ (above) and let A be a nonempty subset of Γ. If b is any element of A we define

$$A \operatorname{rep} b = \{aib \mid i = 0, 1, 2, \cdots, a \in A \backslash b\}$$

where by definition

$$a \circ b = a, \text{ and inductively, } ai + 1b = (aib)b \ (i \geq 0).$$

Thus $A \operatorname{rep} b$ consists of the elements of A, excluding b, and those obtained from them by repeatedly multiplying them on the right by b thus: $((\cdots(ab)\cdots)b)$. Products in which the bracketing is of this form are called *left-normed*.

Now, returning to G and X again, we are able to describe our second procedure for obtaining a basic sequence. We put $X_1 = X$. Supposing X_n $(n \geq 1)$ has already been defined, we choose any element b_n in X_n of minimal length and put

$$X_{n+1} = X_n \operatorname{rep} b_n.$$

It is easy to prove then that the resultant sequence $b_1, b_2, \cdots$ is a basic sequence in X.

The importance of basic sequences lies in the fact that they enable one to "linearise" certain groups and lie rings.

We recall that a lie ring L is a nonempty set with two binary operations $+$ and $[\ ,\ ]$ such that, relative to $+$, L is an abelian group, relative to $[\ ,\ ]$ we have left and right distributivity, and for all $l, m, n \in L$,

$$[l, l] = 0, \ [l, m, n] + [m, n, l] + [n, l, m] = 0,$$

where $[l, m, n] \equiv [[l, m], n]$. Thus

$$0 = [l + m, l + m] = [l, l] + [l, m] + [m, l] + [m, m] = [l, m] + [m, l],$$

so that

$$[m, l] = - [l, m], \ l, m \in L.$$

If $\Xi = \{\xi_1, \cdots, \xi_q\}$ generates L (i.e. L is the smallest sub-lie ring of L containing Ξ) we write $L = lr(\Xi)$.

Suppose now that L is the lie ring generated by $\Xi = \{\xi_1, \cdots, \xi_q\}$. If we forget about the addition in L, L becomes a groupoid. Let Γ be the subgroupoid of L generated by Ξ and let $\beta_1, \beta_2, \cdots$ be a basic sequence in Ξ. These elements β_1, $\beta_2, \cdots$ are often referred to as *basic products*. These basic products enable us to linearise L (P. Hall [32]):

Theorem 3.5. *Let L be a lie ring generated by $\Xi = \{\xi_1, \cdots, \xi_q\}$ and let β_1, $\beta_2, \cdots$ be any basic sequence in Ξ. Then L is spanned additively by the β_i ($i = 1, 2, \cdots$).*

The proof is easy if we first observe the following

Lemma 3.2. *Let L be a lie ring generated by a (possibly infinite) set Y and let $y \in Y$. If $M = lr(Y \text{ rep } y)$ then $L = gp(y) + M$.*

(Of course, $gp(y)$ stands for the additive group generated by y.)

Proof. We first prove that if $u \in M$ then $[u, y] \in M$. It is enough to prove that for each monomial $m = m(c_1, \cdots, c_n)$ in the elements c_j of Y rep y, $[m, y] \in M$. Now each such monomial m has a formal length $\|m\|$, viz. the number of factors c_j that it is formally composed of. If $\|m\| > 1$ then m is formally (uniquely) a product $m = [m', m'']$ where $m', m'' \in M$ are shorter monomials (in the c_j) than m. Using the Jacobi identity we have

$$[m, y] = [[m', m''], y] = - [[m'', y], m'] + [[m', y], m''].$$

So inductively $[m, y] \in M$.

In order now to prove the lemma, it clearly suffices to show that if $l = l(y, z, \cdots)$ is a Y-monomial with the property that some $z \in Y$, $z \neq y$, appears in l, then $l \in M$. To do this we proceed by induction on $|l|$, the formal length of l as a monomial in the *elements of* Y. If $|l| = 1$ the result is clear. If $|l| > 1$ then $l = [l', l'']$ where $|l'| < |l|$, $|l''| < |l|$. By our hypothesis on l, either l' or l'', say l', contains an element of Y different from y. Hence, inductively, this l' lies in M. Now if $l'' = y$, then $[l', l''] \in M$ by our remarks above concerning M. Otherwise $l'' \in M$ by induction also, and hence so does $[l, l'']$. This proves the lemma.

We proceed now to the proof of Theorem 3.5. To this end we define

$$\Xi_1 = \Xi, \ldots, \Xi_{n+1} = \Xi_n \operatorname{rep} \beta_n, \ldots .$$

By Lemma 3.2

$$\begin{aligned} L &= lr(\Xi_1) \\ &= gp(\beta_1) + lr(\Xi_2) \\ &= gp(\beta_1) + gp(\beta_2) + lr(\Xi_3) \\ &= \cdots . \end{aligned} \tag{1}$$

But notice that if m is any monomial in $\xi_1, \ldots, \xi_q$ of (formal) length n and if r is chosen so that every element of X_{r+1} is of (formal) length at least $n+1$, then

$$m \in gp(\beta_1) + \cdots + gp(\beta_r). \tag{2}$$

For in the expansions obtained in (1) the monomials in $\xi_1, \cdots, \xi_q$ are always re-expressed as sums of monomials, each of which has the same length as the original ones (see the proof of Lemma 3.2). The equation (2) follows immediately from this remark. So L is additively generated by the elements $\beta_1, \beta_2, \cdots$.

To see how these notions relate to group theory we recall that to each group G one can associate a lie ring L in the following way. Additively L is the (restricted direct) sum of the groups $\lambda_i G/\lambda_{i+1} G$:

$$L = \coprod_{i=1}^{\infty} \lambda_i G/\lambda_{i+1} G.$$

We define a "lie multiplication" in L by

$$[a\lambda_{i+1} G,\ b\lambda_{j+1} G] = [a, b]\lambda_{i+j+1} G$$

for $a \in \lambda_i G$, $b \in \lambda_j G$. It follows from Corollary 0.31 that this multiplication is unambiguous. Moreover, if we extend it, by distributivity, to the whole of L, then L becomes a lie ring. This idea is due to M. Lazard [53]. Although much of the finer commutator structure in G is lost in L, a great deal has been accomplished by going over to such a lie ring, instead of concentrating on the group G itself. The most significant application is to the restricted Burnside problem: *there is a bound on the orders of all finite n-generator groups of prime exponent* p (A. I. Kostrikin [50]; see G. Higman's discussion in [40]). For a different type of application see G. Higman [39].

Here we need the following observation which is an immediate consequence of the definitions.

Lemma 3.3. *Let G be a group generated by a set X. Then the lie ring L of G is generated by the set* $\{x\lambda_2 G | x \in X\}$.

Suppose now that G is a group generated by the finite set $Y = \{y_1, \cdots, y_q\}$. Disregarding the fact that G is a group, we may turn G into a groupoid by introducing a new binary operation in G, namely commutation [,]:

$$[g, h] = g^{-1}h^{-1}gh \, (g, h \in G).$$

Then Y generates a subgroupoid of the groupoid G. Let

$$c_1, c_2, \cdots \tag{3}$$

be a basic sequence in Y. The c_i are termed *basic commutators in Y*. Each basic commutator c_i has a *weight* associated with it, namely the length of the corresponding element b_i of the free groupoid on X.

Now put $\xi_i = y_i\lambda_2 G$ $(i = 1, \cdots, q)$ and let $\Xi = \{\xi_1, \cdots, \xi_q\}$. The basic lie products in Ξ span L since Ξ generates L (Theorem 3.5). In particular it follows that the basic lie products of length r span $\lambda_r G/\lambda_{r+1} G$. It follows from this remark and the definition of the sequence $c_1, c_2, \cdots$, that we have proved the following theorem of P. Hall [32]:

Theorem 3.6. *Let G be a group generated by $Y = \{y_1, \cdots, y_q\}$ and let c_1, $c_2, \cdots$ be any basic sequence in Y. Then $\lambda_r G$ is generated, modulo $\lambda_{r+1} G$, by the basic commutators of weight r $(r = 1, 2, \cdots)$.*

Corollary 3.61. *Let the group G be generated by the set* $Y = \{y_1, \cdots, y_q\}$. Then *G is nilpotent if and only if all but a finite number of the elements of any basic sequence in Y are equal to* 1.

Corollary 3.61 enables us to prove a theorem of K. W. Gruenberg about Engel groups. A group G is an *Engel group* if for every pair g, h of elements of G there exists an integer n such that $[g, nh] = 1$, where $[g, (i+1)h] = [[g, ih], h]$. The stimulus for the study of Engel groups was provided initially by the theory of lie rings (see, e.g., [46]). It was hoped at one time that a finitely generated Engel group G would turn out to be nilpotent. This is so if G is finite (H. Zassenhaus [88], M. Zorn [89]). However, E. S. Golod and I. R. Šafarevič proved this false in general [23]. There are many interesting results in this area (see, e.g., [6], [26], [36]). As we remarked above, our objective is to prove the following theorem of K. W. Gruenberg [24].

Theorem 3.7. *A finitely generated solvable Engel group G is nilpotent.*

Proof. We use Corollary 3.61. Suppose $G = gp(Y)$ where $Y = \{y_1, \cdots, y_q\}$. Let $c_1, c_2, \cdots$ be a basic sequence in Y, and put

$$Y_1 = Y,\ Y_2 = Y_1 \operatorname{rep} c_1, \cdots, Y_{n+1} = Y_n \operatorname{rep} c_n, \cdots .$$

Since G is an Engel group each of the sets Y_n contains only finitely many elements $\neq 1$. Moreover M_{q+1} clearly consists of elements in $\delta_1 G$. Similarly if l is chosen sufficiently large, Y_m consists of elements in $\delta_2 G$ whenever $m \geq l$. Inductively, for every $k \geq 1$ there exists an integer $r(k)$ such that Y_n consists of elements in $\delta_k G$ provided $n \geq r(k)$. Since G is solvable this leads to the desired conclusion that each of the sets Y_n consists of only the identity element whenever n is sufficiently large. So by Corollary 3.61, G is nilpotent. This completes the proof of the theorem.

3.4. We proceed with our discussion of basic commutators as they relate to free groups. In this section by "ring", without qualification, we shall always mean associative ring with 1. Suppose R is such a ring. Then R can be made into a lie ring by introducing a new binary operation (,) in R termed ring *commutation*:

$$(a, b) = ab - ba, \quad (a, b \in R).$$

Under the operations of addition and commutation, R is a lie ring, called the *commutation lie ring on* R.

We shall be concerned with a lie subring of the commutation lie ring of a free ring. Suppose that R is the free (associative) ring freely generated by

$$Y = \{y\} \cup \{z_\lambda | \lambda \in \Lambda\}$$

and let Δ be the lie subring of the commutation lie ring on R generated by Y. We know from Lemma 3.2 that $\Delta = gp(y) + lr(Y \operatorname{rep} y)$. We are interested in the subring (i.e. sub-associative-ring) of R generated by $Y \operatorname{rep} y$; the information we shall obtain about this subring will enable us to prove that certain basic commutators are linearly independent. Indeed we now prove (see [61], p. 319).

Lemma 3.4. *Let R be the free ring freely generated by*

$$Y = \{y\} \cup \{z_\lambda | \lambda \in \Lambda\}$$

and let S be the subring generated by $Y \operatorname{rep} y$. Then S is a free ring freely generated by $Y \operatorname{rep} y$.

Proof. Put

$$z_{\lambda,i} = (z_\lambda, \underbrace{y, \cdots, y}_{i}) \quad (i = 0, 1, \cdots, \lambda \in \Lambda).$$

Thus $z_{\lambda,0} = z_\lambda$. The elements of $S = rg(Y \operatorname{rep} y)$ are simply linear combinations of monomials in the $z_{\lambda,i}$. It is worthwhile to compute $z_{\lambda,i}$:

$$z_{\lambda,i} = z_\lambda y^i - \binom{i}{1} y z_\lambda y^{i-1} + \binom{i}{2} y^2 z_\lambda y^{i-2} - \cdots + (-1)^i y^i z_\lambda .$$

Notice that each $z_{\lambda,i}$ is a homogeneous element of degree $i+1$. Hence a monomial

$$z_{\lambda_1,i_1} z_{\lambda_2,i_2} \cdots z_{\lambda_k,i_k} \tag{1}$$

(in Y rep y) will be homogeneous of degree

$$(i_1+1)+(i_2+1)+\cdots+(i_k+1).$$

Now in order to prove that S is free on Y rep y it is enough to prove that the formally distinct monomials in Y rep y are (additively) linearly independent. Since, as we observed above, these monomials are homogeneous, it is enough to prove that the monomials of the same (apparent) degree are linearly independent. To do this we impose a total order on Y with y the first element in this ordering. This gives rise to a lexicographic ordering of the monomials in Y of the same degree. It follows without difficulty that the "last" monomial in the expansion of (1) is

$$z_{\lambda_1} y^{i_1} z_{\lambda_2} y^{i_2} \cdots z_{\lambda_k} y^{i_k}.$$

It is then clear that this last monomial determines the Y rep y monomial (1) uniquely. Consequently any nontrivial linear combination of distinct Y rep y monials of the same Y-degree will be nonzero. This completes the proof of the lemma.

Lemma 3.4 makes it easy for us to prove

Theorem 3.8. *Let R be a free ring freely generated by $Y = \{y_1, \cdots, y_q\}$. Further, let Λ be the lie subring of the commutation lie ring on R generated by Y. Then the terms of any basic sequence in Y are additively linearly independent.*

Proof. Let $b_1, b_2, \cdots$ be a basic sequence in Y and put

$$Y_1 = Y,\ Y_2 = Y \operatorname{rep} b_1, \cdots, Y_{n+1} = Y_n \operatorname{rep} b_n, \cdots$$

where we emphasize that the binary operation involved here is ring commutation.

We know, by Theorem 3.5, that the b_i span Λ (additively). Here we are trying to establish their independence. Now by Lemma 3.2 $\Lambda = lr(b_1) + lr(Y_2)$. Our aim is to prove no nonzero multiple of b_1 lies in $lr(Y_2)$. To this end let I be the ideal of the *ring* R generated by $b_2, \cdots, b_q$. Then of course $lr(Y_2) \subseteq I$. But R/I is a polynomial ring in $b_1 + I$ and hence a nonzero multiple of b_1 does not even lie in I. This means Λ is the *direct* sum of $lr(b_1)$ and $lr(Y_2)$:

$$\Lambda = lr(b_1) \oplus lr(Y_2).$$

Let now S be the subring of the ring R generated by Y_2. By Lemma 3.4 S is a free ring freely generated by Y_2. It follows therefore, by a repetition of the argument above, that

$$\Lambda = lr(b_1) \oplus \cdots \oplus lr(b_2) \oplus lr(Y_3).$$

Consequently we find, inductively, for every $n \geq 1$,

$$\Lambda = lr(b_1) \oplus \cdots \oplus lr(b_n) \oplus lr(Y_{n+1}).$$

This completes the proof of the theorem.

Corollary 3.81. *Let R be a free ring freely generated by $Y = \{y_1, \cdots, y_q\}$ and let Λ be the lie subring of the commutation lie ring on R generated by Y. Then Λ is a free lie ring freely generated by Y.*

Proof. Let L be the free lie ring freely generated by $\Xi = \{\xi_1, \cdots, \xi_q\}$ and let ϕ be a homomorphism of L onto Λ defined by ϕ: $\xi_i \to y_i$ $(i = 1, \cdots, q)$. Let $\beta_1, \beta_2, \cdots$ be a basic sequence in Ξ. Then $\beta_1\phi, \beta_2\phi, \cdots$ is a basic sequence in Y. By Theorem 3.5 the β_j span L. Thus if $a \in L$ then a can be expressed in the form

$$a = m_1\beta_1 + \cdots + m_k\beta_k.$$

If $a \neq 0$ then at least one $m_j \neq 0$. So, by the linear independence of the $\beta_i\phi$ we find

$$m_1(\beta_1\phi) + \cdots + m_k(\beta_k\phi) \ (= a\phi) \neq 0.$$

It follows that ϕ is an isomorphism, as desired.

Theorem 3.8 enables one to prove easily a number of results involving the lower central series of a free group. First we prove (cf. Theorem 3.1)

Lemma 3.5. *Let F be the free group on $X = \{x_1, \cdots, x_q\}$, let R be the free ring on $Y = \{y_1, \cdots, y_q\}$, and let Λ be the lie subring of the commutation lie ring on R generated by Y. Let Ψ be the Magnus embedding of F in $\bar{R}$ given by*

$$\Psi: x_i \to 1 + y_i, \ 1 \leq i \leq q.$$

Let $b_1, b_2, \cdots$ be any basic sequence in X, $\tilde{b}_1, \tilde{b}_2, \cdots$ the corresponding sequence of basic lie products in Λ. Then

$$\Psi: b_n \to 1 + \tilde{b}_n + \cdots \ (n = 1, 2, \cdots)$$

(where "$\cdots$" here and in the proof stands for "terms of higher degree").

Proof. We proceed by induction on $|b_n|$, the result being clear if $|b_n| = 1$. Suppose $|b_n| > 1$, then b_n can be expressed uniquely as $b_n = [b_i, b_j]$ where $|b_i| < |b_n|$, $|b_j| < |b_n|$. Therefore, inductively

$$\begin{aligned}
b_n\Psi &= [b_i, b_j]\Psi \\
&= (1 + \tilde{b}_i + \cdots)^{-1}(1 + \tilde{b}_j + \cdots)^{-1}(1 + \tilde{b}_i + \cdots)\,(1 + \tilde{b}_j + \cdots) \\
&= (1 + \tilde{b}_i + \cdots)^{-1}(1 + \tilde{b}_j + \cdots)^{-1}\{(1 + \tilde{b}_j + \cdots)(1 + \tilde{b}_i + \cdots) \\
&\quad + (1 + \tilde{b}_i + \cdots)(1 + \tilde{b}_j + \cdots) - (1 + \tilde{b}_j + \cdots)(1 + \tilde{b}_i + \cdots)\} \\
&= 1 + (1 + \tilde{b}_i + \cdots)^{-1}(1 + \tilde{b}_j + \cdots)^{-1}\{\tilde{b}_i\tilde{b}_j - \tilde{b}_j\tilde{b}_i + \cdots\} \\
&= 1 + (\tilde{b}_i\tilde{b}_j - \tilde{b}_j\tilde{b}_i) + \cdots \\
&= 1 + \tilde{b}_n + \cdots .
\end{aligned}$$

Thus the lemma follows.

Theorem 3.9. *Let F be the free group on $X = \{x_1, \cdots, x_q\}$, let $b_1, b_2, \cdots$ be a basic sequence in X and let r be a positive integer. Then modulo $\lambda_{r+1}F$, $\lambda_r F$ is a free abelian group freely generated by the basic commutators b_j of weight r.*

Proof. We have already shown, in Theorem 3.6, that the basic commutators $b_l, \cdots, b_m$ of weight r generate $\lambda_r F$ modulo $\lambda_{r+1}F$. So it only remains to prove that they are independent modulo $\lambda_{r+1}F$.

Using the notation of the previous lemma, we have that for any integers $k_l, \cdots, k_m$,

$$\Psi \colon b_l^{k_l} \cdots b_m^{k_m} \longrightarrow 1 + (k_l\tilde{b}_l + \cdots + k_m\tilde{b}_m) + \cdots .$$

Since, by Theorem 3.8, the basic lie elements $\tilde{b}_l, \cdots, \tilde{b}_m$ are additively linearly independent, it follows at once that $b_l, \cdots, b_m$ are independent modulo $\lambda_{r+1}F$. This completes the proof of the theorem.

It is now an easy matter to obtain the precise rank of $\lambda_r F/\lambda_{r+1}F$ where F is a free group on $x_1, \cdots, x_q$; we must simply compute the number of basic commutators in F of weight r (see, e.g., [61], p. 330). More important perhaps are the following corollaries of Theorem 3.9.

Corollary 3.91. *Let F be the free group on $X = \{x_1, \cdots, x_q\}$. Then the lie ring of F is a free lie ring freely generated by q elements.*

Proof. This could be proved in several ways using the information we now have at hand. For example, let L be the lie ring of F; then L is generated by the set $X' = \{x_1\lambda_2 F, \cdots, x_q\lambda_2 F\}$. By Theorem 3.9, any sequence of basic lie elements (in X') is a linearly independent set. Corollary 3.91 now follows easily by inspecting the proof of Corollary 3.81.

In order to explain the second corollary, let G be any group and let ZG denote the group ring of G over the ring of integers Z. Let $A(G)$ denote the augmentation ideal of ZG, i.e., the ideal of all elements of ZG whose coefficient sum is zero (see 4.4). For n a positive integer let $(A(G))^n$ be the two-sided ideal of ZG generated by all elements of the form $a_1 a_2 \cdots a_n$, $a_i \in A(G)$, $1 \leq i \leq n$. Then

$$d_n G = (1 + A(G)^n) \cap G$$

is termed the *nth dimension subgroup of* G. An easy consequence of Theorem 3.9 is the following:

Corollary 3.92. *Let* F *be a free group. Then* $d_n F = \lambda_n F$ *for* $n = 1, 2, \cdots$.

It is an unsolved problem whether $d_n G = \lambda_n G$ for every group G (see, e.g., [76] for further details). It is, incidentally, easy to prove that $d_n G \geq \lambda_n G$ for every n.

Chapter 4. Lie group techniques

4.1. The interrelationship between lie groups and lie algebras has many important applications. In this chapter we shall establish a similar relationship between $\mathfrak{D}$-groups and rational nilpotent lie algebras discovered by A. I. Mal'cev [62]. The approach we shall use is based on the work of S. A. Jennings [47], centering on group algebras and the Baker-Campbell-Hausdorff formula. This leads to an equivalence between the category of nilpotent $\mathfrak{D}$-groups and the category of nilpotent lie algebras over the field Q of rationals. This equivalence gives rise to a connection between the automorphism groups of finitely generated nilpotent groups and arithmetic groups; and this leads to a proof (which we shall give in 4.6) that the automorphism group of a finitely generated nilpotent group is finitely presented (L. Auslander and G. Baumslag [4]). Recently L. Auslander [3] has established, by the same basic approach, that the automorphism group of *every* polycyclic group is finitely presented. It seems likely that this technique linking finitely generated nilpotent groups to arithmetic groups will be useful in other contexts.

4.2. Our main concern in this chpater is to put ourselves in a position be be able to find the automorphism group of a finitely generated nilpotent group. W. Gaschutz [22] has shown that every finite p-group G $(\neq 1)$ has an outer automorphism. Here by imitating the proof of the corresponding theorem for outer derivations of a finite dimensional nilpotent lie algebra due to Nathan Jacobson [46] we shall prove

Theorem 4.1. *A finitely generated nilpotent group G has outer automorphisms of infinite order if and only if G is of torsion-free rank at least two.*

We recall that the group of outer automorphisms of G is the factor group of the automorphism group of G by the group of inner automorphisms of G. It is easy to see that if G is a finitely generated nilpotent group of torsion-free rank at most 1 then the outer automorphism group of G is finite. So the burden of the proof of Theorem 4.1 is to show that if G is a finitely generated nilpotent group of torsion-free rank at least two, then the outer automorphism group of G has elements of infinite order. The rest of the proof is most conveniently carried out in two parts. First we suppose that G is torsion-free. We proceed in this case as follows. For $i = 1, 2, \cdots$, define

$$\bar{\lambda}_i G = \{x \in G \mid x^n \in \lambda_i G \text{ for some } n \geq 1\}.$$

It is easy to see that $\bar{\lambda}_i G$ is a normal subgroup of G for every i. Now choose a normal subgroup F of G such that G/F is infinite cyclic, and choose $a \in G$ so that

$G = gp(a, F)$. Let $Z = \zeta F$. Then since $\overline{\lambda}_{c+1} G = 1$ we can find an integer $n \geq 1$ such that

$$Z \leq \overline{\lambda}_n G,\ Z \not\leq \overline{\lambda}_{n+1} G.$$

Choose $z \in Z$, $z \notin \overline{\lambda}_{n+1} G$. We define a mapping $\theta\colon G \to G$ by

$$\theta\colon a^l f \to (az)^l f \ (f \in F,\ l \text{ any integer}).$$

Then

$$(a^l f a^{l'} f')\theta = (a^{l+l'} f^{a^{l'}} f')\theta = (az)^{l+l'} f^{a^{l'}} f',$$

and

$$(a^l f)\theta (a^{l'} f')\theta = (az)^l f \cdot (az)^{l'} f' = (az)^{l+l'} f^{a^{l'}} f'.$$

So θ is an endomorphism. Clearly we can replace z by z^{-1} throughout which shows θ has an inverse and hence θ is an automorphism of G.

We claim that none of the powers $\phi = \theta^t$ $(t \neq 0)$ of θ is an inner automorphism of G. Suppose if possible that ϕ is the inner automorphism of G induced by h:

$$\phi\colon g \to h^{-1} g h \ (g \in G).$$

Let us write $h = a^l f$, where $f \in F$. Then

$$a\phi = f^{-1} a f = a[a, f].$$

But $a\phi = az^t$ $(t \neq 0)$. So

$$[a, f] = z^t. \tag{1}$$

Now h must centralize F since ϕ leaves F identically fixed. Therefore, in particular,

$$1 = [h^{-1}, f] = a^l f f^{-1} f^{-1} a^{-l} f.$$

So a^l centralises f. It follows that

$$(f^{-1} a f)^l = f^{-1} a^l f = a^l.$$

But G is a $\mathfrak{U}$-group. Therefore either $l = 0$ or $f^{-1} a f = a$. This however would contradict (1). Therefore $l = 0$ and $h \in F$. But again we observe that ϕ leaves F identically fixed. So $h \in Z$. Consequently $h \in \overline{\lambda}_n G$ which means that, by (1), $z^t \in \overline{\lambda}_{n+1} G$. So, by the very definition of $\overline{\lambda}_{n+1} G$ we have $z \in \overline{\lambda}_{n+1} G$. This contradiction completes the proof that no nontrivial power of G is an inner automorphism of G. This completes the proof of the first part of Theorem 4.1.

In order to prove the other part of Theorem 4.1, i.e., that *every* finitely generated nilpotent group of torsion-free rank at least two has an outer automorphism of infinite order, embed G as a subgroup of finite index in a direct product $P = F \times T$, where F is finite and T is a finitely generated torsion-free nilpotent group. It is easy to see that T is of torsion-free rank at least two. So T has an automorphism θ none of whose nontrivial powers is an inner automorphism of T. It follows that if θ^+ is the automorphism of P which is the identity on F and θ on T, then θ^+ gives rise to an element of infinite order in the outer automorphism group of P. Since G is of finite index in P some nontrivial power $(\theta^+)^l$ of θ^+ induces an automorphism in G itself (see Lemma 4.3). This is the desired automorphism of G. For suppose $(\theta^+)^{lm}$ $(lm \neq 0)$ is an inner automorphism of G. Then there exist elements $u \in T$, $v \in F$ such that $(\theta^+)^{lm}$ acts on G by conjugation with vu. Thus θ^{lm} induces on $H = G \cap T$ the automorphism obtained by conjugating by u. Let μ denote the inner automorphism of T obtained by conjugating by u. Then $\theta^{lm}\mu^{-1} = \Psi$ is an automorphism of the torsion-free nilpotent group T which leaves a subgroup H of finite index identically fixed. Now the Mal'cev completions $m(T)$, $m(H)$ of T and H (respectively) coincide. It follows that $m(\psi)$ and $m(\iota)$ coincide too, where ι is the identity automorphism of H. Hence ψ is the identity automorphism of T and θ^{lm} is inner. This contradiction completes the proof of Theorem 4.1.

4.3. We begin our lie-theoretic approach to automorphism groups by discussing group algebras.

Let R be a ring and let G be a group. We denote the group algebra of G over R by RG. Thus RG is the set of all functions from G into R which have finite support. RG is turned into an R-algebra by defining, for $f_1, f_2 \in RG$ and $r \in R$,

$$(f_1 + f_2)(a) = f_1(a) + f_2(a),$$
$$(f_1 \cdot f_2)(a) = \sum_{x \in G} f_1(ax) f_2(x^{-1}),$$
$$(rf_1)(a) = rf_1(a).$$

As usual we denote the elements of RG by finite sums of multiples of elements of G by elements of R.

Following G. Higman [37] we term a group *locally indicable* if every nontrivial finitely generated subgroup can be mapped homomorphically onto the infinite cyclic group. We then have the following theorem of G. Higman [37].

Theorem 4.2. *The group algebra of a locally indicable group G over an integral domain R is an integral domain.*

Here we are using "integral domain" in the noncommutative sense, i.e. an *integral domain* is a ring with 1 which contains no proper zero divisors.

Proof. Let $f, g \in RG$. On multiplying f and g by elements of G if necessary, we may assume

$$f = r_0 + r_1 a_1 + \cdots + r_m a_m,$$
$$g = s_0 + s_1 b_1 + \cdots + s_n b_n,$$

$(r_i, s_j \in R,\ a_i, b_j \in G,\ 1 \leq i \leq m,\ 1 \leq j \leq n)$. We have to show that if $f \neq 0$, $g \neq 0$, then $fg \neq 0$. Suppose that the elements $r_0, \cdots, r_m, s_0, \cdots, s_n$ are nonzero. The proof is by induction on $m + n$. If $m + n = 1$ the result is clear. So assume $m + n > 1$, and let

$$H = gp(a_1, \cdots, a_m, b_1, \cdots, b_n).$$

Choose K normal in H such that H/K is infinite cyclic on xK. Then we can write, for all relevant values of i and j,

$$a_i = x^{\alpha_i} a_i', \ b_j = x^{\beta_j} b_j' \ (a_i', b_j' \in K).$$

Thus we find

$$f = r_0 + x^{\gamma_1} f_1 + \cdots + x^{\gamma_k} f_k \ (\gamma_1 < \gamma_2 < \cdots < \gamma_k),$$
$$g = s_0 + x^{\delta_1} g_1 + \cdots + x^{\delta_l} g_l \ (\delta_1 < \delta_2 < \cdots < \delta_l),$$

where the f_i and g_j are in RK and each involves fewer summands than m and n respectively. Note that at least one of the exponents $\gamma_1, \cdots, \gamma_k, \delta_1, \cdots, \delta_l$ must be nonzero because $K \neq H$. Now

$$fg = r_0 s_0 + \cdots + x^{\gamma_k} f_k x^{\delta_l} g_l.$$

Choose γ_i and δ_j so that $\gamma_i + \delta_j$ is maximum among the sums $\gamma_s + \delta_t$ $(0 \leq s \leq k$, $0 \leq t \leq l$ where by definition $\gamma_0 = 0 = \delta_0)$. Inductively

$$x^{\gamma_i} f_i x^{\delta_j} g_j \neq 0.$$

Moreover there is no term in fg which can cancel with $x^{\gamma_i} f_i x^{\delta_j} g_j$ by our choice of $\gamma_i + \delta_j$. It follows that $fg \neq 0$. This completes the proof of Theorem 4.2.

4.4. Our present objective is to establish a connection between nilpotent $\mathfrak{D}$-groups and nilpotent lie algebras. The basic result needed in our development is a theorem of S. A. Jennings [47]. It should be pointed out, however, that only by going over to the theory of lie groups can one establish the results in sufficient generality to be able to derive the full benefit from the relationship between groups and lie algebras.

In order to explain this theorem of Jennings, let G be any group, R any ring with 1. The *augmentation ideal* $A(R, G)$ of RG is as we remarked earlier, the kernel of the augmentation map of RG onto R given by

$$\sum_{i=1}^{m} r_i g_i \longrightarrow \sum_{i=1}^{m} r_i.$$

We shall usually suppress reference to R and G and simply denote the augmentation ideal of RG by A.

RG is termed *residually nilpotent* if

$$\bigcap_{n=1}^{\infty} A^n = (0),$$

where here A^n is the two-sided ideal of RG generated by all elements of the form $a_1 a_2 \cdots a_n$, $a_i \in A$, $1 \leq i \leq n$.

Theorem 4.3. (S. A. Jennings [47]). *Let G be a finitely generated torsion-free nilpotent group and let F be a field. Then FG is residually nilpotent.*

The restriction that G be finitely generated is unnecessary as M. Lazard (unpublished) has shown. We shall concern ourselves here mainly with this case.

Let

$$G = G_0 > G_1 > \cdots > G_r = 1$$

be a central series for G in which G_i/G_{i+1} is infinite cyclic, say on $a_{i+1}G_{i+1}$, $0 \leq i \leq r-1$. Then

Lemma 4.1. *The formally distinct products of the form*

$$(a_{i_1}^{\pm 1} - 1)^{\epsilon_1} \cdots (a_{i_s}^{\pm 1} - 1)^{\epsilon_s} \quad (1 \leq i_1 < i_2 < \cdots < i_s \leq r,\ \epsilon_j > 0) \tag{1}$$

form a basis for the vector space A over F.

Proof. First observe that the elements $g - 1$ ($g \in G$, $g \neq 1$) constitute a basis for A. Next we note that

$$xy - 1 = (x-1)(y-1) + (x-1) + (y-1). \tag{2}$$

Thus every element of the form

$$(a_{i_1}^{\pm 1})^{\epsilon_1} \cdots (a_{i_s}^{\pm 1})^{\epsilon_s} - 1 \quad (i_1 < i_2 < \cdots < i_s)$$

can be expressed as a sum of products of the $(a_{i_j}^{\pm 1} - 1)$'s (in the right order). As every element $g \in G$ is of the form $g = (a_{i_1}^{\pm 1})^{\epsilon_1} \cdots (a_{i_s}^{\pm 1})^{\epsilon_s}$, the identity (2) leads to an expression for each $g-1$ $(g \in G)$ as a sum of elements of the form (1). Thus, by our first remark, the elements of the form (1) contain a basis for A.

To see that these elements are independent over F, note that two distinct elements of the form

$$(a_{i_1}^{\pm 1}-1)^{\epsilon_1} \cdots (a_{i_s}^{\pm 1}-1)^{\epsilon_s} \quad (1 \le i_1 < i_2 < \cdots < i_s \le r,\ \epsilon_j > 0)$$

have distinct "leading terms" $a_{i_1}^{\pm\epsilon_1} \cdots a_{i_s}^{\pm\epsilon_s}$. Therefore it follows that a nontrivial F-linear sum of elements of the form (1) cannot be zero. This completes the proof of the lemma.

Consider now an arbitrary product of the form

$$p = (a_{l_1}^{\pm 1}-1)(a_{l_2}^{\pm 1}-1) \cdots (a_{l_n}^{\pm 1}-1) \tag{3}$$

where each $l_j \in \{1, \cdots, r\}$ but otherwise there is no restriction; in particular both $(a_s - 1)$ and $(a_s^{-1}-1)$ may occur as factors. We define the *weight* $\mathrm{wt}(p)$ of this product p by

$$\mathrm{wt}(p) = 2^{l_1} + 2^{l_2} + \cdots + 2^{l_n}.$$

Let us define a *straight product* to be a product of the form

$$\prod_{i=1}^{r} (a_i^{-1}-1)^{\alpha_i}(a_i - 1)^{\beta_i}) \tag{4}$$

where α_i, β_i are nonnegative integers, $1 \le i \le r$. Note that the weight of this straight product (4) is $\Sigma_{i=1}^{r} (\alpha_i + \beta_i)2^i$. We then have

Lemma 4.2. *Every product of the form* (3) *can be expressed as a linear combination of straight products of no smaller weight.*

Proof. First note the identity

$$\begin{aligned}(y-1)(x-1) &= (x-1)(y-1) + ([y,x]-1) + (x-1)([y,x]-1) \\ &\quad + (y-1)([y,x]-1) + (x-1)(y-1)([y,x]-1).\end{aligned} \tag{5}$$

Now if $s < t$,

$$[a_t^{\pm 1}, a_s^{\pm 1}] = (a_{t+1}^{\pm 1})^{\nu_{t+1}} \cdots (a_r^{\pm 1})^{\nu_r},$$

for suitable nonnegative integers $\nu_{t+1}, \cdots, \nu_r$. Therefore, using the identity (2), we have

$$[a_t^{\pm 1}, a_s^{\pm 1}] - 1 = \Sigma \rho q$$

where the ρ are integers and the q are straight products involving only $a_{t+1}, \cdots, a_r$. Clearly $\mathrm{wt}(q) \geq 2^{t+1}$ for all q occurring. Moreover

$$\mathrm{wt}((a_t^{\pm 1} - 1)(a_s^{\pm 1} - 1)) = 2^t + 2^s < 2^{t+1}.$$

It now follows from the identity (5) that

$$(a_t^{\pm 1} - 1)(a_s^{\pm 1} - 1) = (a_s^{\pm 1} - 1)(a_t^{\pm 1} - 1) + \Sigma \rho' q',$$

where the ρ' are integers and the q' are straight products satisfying

$$\mathrm{wt}(q') > 2^s + 2^t.$$

This completes the proof of Lemma 4.2.

We are now in a position to prove Theorem 4.3. We proceed by induction on the torsion-free rank r of G. If $r = 1$, G is infinite cyclic so that FG is a principal ideal domain. In this case every element of A^n is a product of at least n primes. Since FG is a unique factorization domain $\bigcap_{i=1}^{\infty} A^n = (0)$.

Suppose now that $r > 1$ and that the desired conclusion holds for all finitely generated torsion-free nilpotent groups of rank less than r. Suppose that

$$x \in \bigcap_{n=1}^{\infty} A^n, \ x \neq 0.$$

By multiplying x by a sufficiently large positive power of a_1, we may assume that if $x = \Sigma r_j g_j$, then each group element g_j when expressed in canonical form as a product of powers of the elements a_i involves only positive powers of a_1. By Lemma 4.1, we can write x uniquely in the form

$$x = \Sigma(\alpha \Pi (a_{i_j}^{\pm 1} - 1)^{\epsilon_j} \ (\epsilon_j > 0)$$

where the α are elements of F, and in each product Π the terms $(a_{i_j}^{\pm 1} - 1)$ occur in their natural order. It is clear, after a glance at the proof of Lemma 4.1, that only powers of $(a_1^{+1} - 1)$ arise in this expression for x (never powers of $(a_1^{-1} - 1)$). Thus we may, by collecting the "coefficients" of the powers of $(a_1 - 1)$ together, rewrite x as

$$x = x_0 + (a_1 - 1)^{m_1} x_1 + \cdots + (a_1 - 1)^{m_n} x_n \quad (0 < m_1 < \cdots < m_n);$$

here each of $x_0, x_1, \cdots, x_n$ is an F-linear sum of products of the form

$$(a_{i_1}^{\pm 1} - 1)^{\delta_1} \cdots (a_{i_s}^{\pm 1} - 1)^{\delta_s} \quad (2 \leq i_1 < i_2 < \cdots < i_s \leq r, \ \delta_j \geq 0).$$

It should be noted that for each i

$$x_i = f_i + \tilde{a}_i \ (f_i \in F)$$

where $\tilde{a}_i \in A_1$ the augmentation ideal of $G_1 = gp(a_2, \cdots, a_r)$.

Let η be the homomorphism of G onto $H = gp(a_1)$ defined by

$$a_1\eta = a_1, \ a_i\eta = 1 \text{ if } i > 1.$$

Then η induces a homomorphism $\hat{\eta}$ (say) of FG onto FH. Now

$$x\hat{\eta} = f_0 + (a_1 - 1)^{m_1} f_1 + \cdots + (a_1 - 1)^{m_n} f_n \ (0 < m_1 < \cdots < m_n).$$

But the intersection of the powers of the agumentation ideal of the infinite cyclic group H is zero. Hence $x\hat{\eta} = 0$. But the elements $(a_1 - 1)^{m_1}, \cdots, (a_1 - 1)^{m_n}$ $(0 < m_1 < \cdots < m_n)$ are linearly independent over F. So $f_0 = \cdots = f_n = 0$ and hence

$$x_i \in A_1 \ (i = 0, 1, \cdots, n).$$

Now by the induction assumption FG_1 is residually nilpotent. Thus if $x_0 \neq 0$, then we may choose k_0 so that $x_0 \notin A_1^{k_0}$. If $x_0 = 0$, then we may assume $x_1 \neq 0$ and so we can choose k_1 so that $x_1 \notin A_1^{k_1}$. Put

$$\begin{aligned} k &= k_0 & &\text{if } x_0 \neq 0, \\ &= k_1 + m_1 & &\text{if } x_0 = 0. \end{aligned}$$

Finally we put $w = 2^r k$.

Observe now that $x \in A^w$. It follows from Lemma 4.1 that x can be expressed as an F-linear sum of products of the form

$$(a_{l_1}^{\pm 1} - 1)(a_{l_2}^{\pm 1} - 1) \cdots (a_{l_u}^{\pm 1} - 1),$$

where the number of factors in these products is at least w. Notice that these products will in general not be straight; however, by Lemma 4.2, each product can be straightened without loss of weight. It follows therefore that x can be written in the form

$$x = \gamma_1 p_1 + \cdots + \gamma_t p_t \ (\gamma_j \in F),$$

where p_j is a straight product

$$\prod_{i=1}^{r} (a_i^{-1} - 1)^{\alpha_i} (a_i - 1)^{\beta_i} \tag{4}$$

of weight at least w. Thus in every straight product occurring we have

$$(\alpha_1 + \beta_1)2^1 + (\alpha_2 + \beta_2)2^2 + \cdots + (\alpha_r + \beta_r)2^r \geq 2^r k.$$

Therefore

$$(\alpha_1 + \beta_1) + (\alpha_2 + \beta_2) + \cdots + (\alpha_r + \beta_r) \geq k.$$

In other words, each of the straightened products p_j is a product of at least k factors. Suppose that these straightened products $p_1, \cdots, p_t$ are numbered in such a way that

$$\text{if } 1 \leq j \leq s, \ p_j \text{ is of the form (4), with } \alpha_1 = \beta_1 = 0,$$

and

$$\text{if } s < j \leq t, \ p_j \text{ is of the form (4), with either } \alpha_1 \neq 0 \text{ or } \beta_1 \neq 0.$$

Equating the two expressions we have obtained for x, we find

$$x_0 + (a_1 - 1)^{m_1} x_1 + \cdots + (a_1 - 1)^{m_n} x_n = \gamma_1 p_1 + \cdots + \gamma_s p_s + \gamma_{s+1} p_{s+1} + \cdots + \gamma_t p_t,$$

where $x_0, x_1, \cdots, x_n, p_1, \cdots, p_s$ belong to A_1. It follows from Lemma 4.1 on noting $(a_1 - 1)(a_1^{-1} - 1) = -(a_1 - 1) - (a_1^{-1} - 1)$, that

$$x_0 = \gamma_1 p_1 + \cdots + \gamma_s p_s, \tag{5}$$

and

$$(a_1 - 1)^{m_1} x_1 + \cdots + (a_1 - 1)^{m_n} x_n = \alpha_{s+1} p_{s+1} + \cdots + \alpha_t p_t. \tag{6}$$

There are now two cases to consider depending on whether $x_0 \neq 0$ or $x_0 = 0$; we shall prove that either case leads to a contradiction. If $x_0 \neq 0$, we consider equation (5). Each of the straight products $p_1, \cdots, p_s$ is a product of at least k factors of the form $(a_i^{\pm 1} - 1)$, $2 \leq i \leq r$. Thus $x_0 \in A_1^k = A_1^{k_0}$, a contradiction. On the other hand, if $x_0 = 0$, we consider equation (6). Note that each straight product p_j, $(s + 1 \leq j \leq t)$ is of the form

$$\begin{aligned} p_j &= (a_1^{-1} - 1)^{\lambda_j} (a_1 - 1)^{\mu_j} q_j \ (q_j \in A_1), \\ &= (a_1 - 1)^{\lambda_j + \mu_j} (-a_1)^{-\lambda_j} q_j. \end{aligned}$$

If for any j with $s + 1 \leq j \leq t$, $\lambda_j + \mu_j < m_1$, then (using the fact that FG is an integral domain) we can cancel the factor $(a_1 - 1)^{\lambda + \mu}$, where $\lambda + \mu = \min(\lambda_j + \mu_j)$, right across equation (6). The resultant equality contradicts Lemma 4.1. Therefore we must have

$$\lambda_j + \mu_j \geq m_1, \ (s + 1 \leq j \leq t).$$

Now cancelling $(a_1 - 1)^{m_1}$ across equation (6) we obtain an expression for x_1 as a sum of products of elements $(a_i^{\pm 1} - 1)$, $2 \leq i \leq r$, each product having at least $k - m_1$ factors. This implies $x_1 \in A_1^{k - m_1} = A_1^{k_1}$, which is again a contradiction. This completes the proof of the theorem.

4.5. Now let us specialize our considerations to the case where F is the field Q of rational numbers. As we observed above QG is residually nilpotent. Let $\overline{QG}$ be the completion of QG with respect to the A-adic topology (see, e.g., [46] for further details). Now if $a \in \bar{A}$ then we define

$$\exp a = 1 + a + \frac{a^2}{2!} + \cdots .$$

Thus $\exp a$ is a well-defined element of $\overline{QG}$. Similarly if $b \in \overline{QG}$ and $b - 1 \in \bar{A}$ then we define

$$\log b = \log(1 + (b-1)) = (b-1) - \frac{(b-1)^2}{2} + \frac{(b-1)^3}{3} - \cdots .$$

Again $\log b$ makes sense in $\overline{QG}$. As usual we find

$$\log \exp a = a \text{ and } \exp \log b = b.$$

Thus exp is a $1-1$ mapping of $\bar{A}$ onto $1 + \bar{A}$ with log a $1-1$ mapping of $1 + \bar{A}$ onto $\bar{A}$ its inverse. The following theorem is crucial in our development.

Theorem 4.4. *The Q-subspace L of $\overline{QG}$ spanned by $\log G$ is a vector space of dimension r, the torsion-free rank of the finitely generated torsion-free nilpotent group G. In addition, L is a nilpotent lie subalgebra of the commutation lie algebra on $\overline{QG}$.*

(We recall that a lie algebra is *nilpotent* if there exists an integer $c \geq 1$ such that the product of any c of its elements taken in any order is zero.) The lie algebra L above is termed *the lie algebra* of G. In some ways L mirrors the properties of G. We shall give only one instance of this relationship between L and G but first we indicate the proof of Theorem 4.4 (S. A. Jennings [47]). In order to do so we need a formula due to Baker, Campbell and Hausdorff (see [46]). In order to explain let us consider the Magnus power series ring $R = Q[[w, x, y]]$ over Q in w, x and y. Then the Baker-Campbell-Hausdorff formula expresses

$$\exp x \cdot \exp y \text{ in the form } \exp z$$

where z is given as a sum

$$z = \sum_{i=0}^{\infty} h_i(x, y)$$

where the ith component $h_i(x, y)$ of z is an element in the commutation lie subalgebra generated by x and y. More precisely

$$h_0(x, y) = 0,\ h_1(x, y) = x + y,\ h_2(x, y) = \tfrac{1}{2}(x, y), \cdots$$

where our notation here is $(x, y) = xy - yx$ (as is usual when we work in a commutation lie algebra). Thus, for each i, we may interpret $h_i(x, y)$ as a word in a *lie algebra*. We shall denote z more appropriately by $x * y$.

Suppose now that Λ is a lie algebra over a field of characteristic zero. Let Λ^n

denote the ideal of Λ generated by all products of at least n factors. We term Λ *residually nilpotent* if $\bigcap_{n=1}^{\infty} \Lambda^n = 0$. A residually nilpotent lie algebra Λ is endowed with a natural topology and as such is a metric space. If Λ is complete in this topology and if the sequence $a_1, a_1 + a_2, \cdots (a_1, a_2, \cdots \in \Lambda)$ converges, then we denote the unique element of Λ to which it converges by $\Sigma_{i=1}^{\infty} a_i$. If $a_i \in \Lambda^{n_i}$ and if the sequence $n_1, n_2, \cdots$ tends to infinity, then $\Sigma_{i=1}^{\infty} a_i$ always exists. It follows therefore that if $a, b \in \Lambda$ then

$$a * b = \sum_{i=0}^{\infty} h_i(a, b)$$

is an element of Λ. Indeed this composition $*$ turns Λ into a group. In order to see this we have to check first that $*$ is an associative multiplication. It was for this reason that we initially introduced R as the Magnus power series ring on three generators. Observe that, of course, $e^w(e^x e^y) = (e^w e^x)e^y$. This implies that

$$h_i(w, \sum_{i=1}^{\infty} h_i(x, y)) = h_i(\sum_{i=1}^{\infty} h_i(w, x), y) \quad (i = 1, 2, \cdots)$$

and hence $w * (x * y) = (w * x) * y$. Now it follows from Theorem 3.8 that if a, b, c are elements of the complete lie algebra Λ above then the mapping

$$w \longrightarrow a, \; x \longrightarrow b, \; y \longrightarrow c$$

can be extended to a homomorphism of the completion (in R) of the commutation lie subalgebra generated by w, x and y. Therefore

$$a * (b * c) = (a * b) * c.$$

So Λ is a semigroup under $*$. But for each $a \in \Lambda$, $a * 0 = a = 0 * a$ and $a * -a = 0 = -a * a$. So Λ is indeed a group under $*$ which we shall henceforth denote by Λ^*. Λ^* will be referred to as the Baker-Campbell-Hausdorff group on Λ. We shall have more to say about it later.

We now indicate how to prove Theorem 4.4. We begin by showing that L generates a nilpotent lie-subalgebra. Suppose, to this end, that $a, b \in G$. Then we can write $a = \exp \alpha$, $b = \exp \beta$. It follows from the Baker-Campbell-Hausdorff formula that

$$\begin{aligned} a^{-1} b^{-1} ab &= \exp(-\alpha) \exp(-\beta) \exp(\alpha) \exp(\beta) \\ &= \exp((\alpha, \beta) + \cdots) \end{aligned}$$

where each of the subsequent terms is a (lie) commutator of higher weight in α and β. Inductively if $g_i = \exp \gamma_i \in G$ $(i = 1, \cdots, n)$, we find

$$[g_1, g_2, \cdots, g_n] = \exp((\gamma_1, \gamma_2, \cdots, \gamma_n) + \cdots).$$

Suppose G is of class c and that $n > c$. Then it follows from the above identity that $(\gamma_1, \gamma_2, \cdots, \gamma_n) + \cdots = 0$. It follows from this argument that if $n > c$ and $g_1, \cdots, g_n$ are any n elements of G and w is any n-fold commutator in $\log g_1, \cdots, \log g_n$, then w can be re-expressed as a sum of commutators in $\log g_1, \cdots, \log g_n$ of weight at least $n+1$. Since $\bigcap_{n=1}^{\infty} A^n = 0$, it follows that $w = 0$. Thus L generates a nilpotent lie subalgebra of the commutation lie subalgebra on $\overline{QG}$.

The second part of the proof is similar and is left to the reader. The point of this part is to verify that the elements $\log a_i$ actually constitute a basis for L over Q and thence that L is indeed a lie subalgebra as required.

Next we observe that, using the notation we have developed, we have

Proposition 4.1. *Let L be the lie algebra of G. Then $\exp L$ is the Mal'cev completion of G.*

Again Proposition 4.1 is easy to prove and we omit the proof here. Notice that we have obtained an alternative proof of the fact that every torsion-free nilpotent group can be embedded in a nilpotent $\mathfrak{D}$-group.

Suppose now that ϕ is an automorphism of G. Then ϕ defines an automorphism of QG and hence an automorphism $\bar{\phi}$ of $\overline{QG}$. Notice that

$$\begin{aligned}(\log g)\bar{\phi} &= ((g-1) - \frac{(g-1)^2}{2} + \cdots)\bar{\phi} \\ &= (g-1)\bar{\phi} - (\frac{(g-1)^2}{2})\bar{\phi} + \cdots \\ &= (g\phi - 1) - \frac{(g\phi-1)^2}{2} + \cdots \\ &= \log(g\phi).\end{aligned}$$

It follows easily that every automorphism of G gives rise to an automorphism of the lie algebra L which maps $\log G$ onto itself.

Conversely suppose ϕ is an automorphism of L which maps $\log G$ onto $\log G$. Then we define a mapping ψ of G into G by

$$g\psi = h \text{ if } (\log g)\phi = \log h \qquad (g \in G).$$

So ψ is a one-to-one mapping of G onto G. In order to see that ψ is an automorphism of G note that L is nilpotent and hence that if $g_1, g_2 \in G$ then $\log g_1 * \log g_2$ makes sense. Thus $(g_1 g_2)\psi = h$ where

$$\log h = (\log(g_1 g_2))\phi = (\log g_1 * \log g_2)\phi = (\log g_1)\phi * (\log g_2)\phi.$$

Therefore, if $(\log g_i)\phi = \log h_i$ $(i = 1, 2)$ we have

$$\log h = \log h_1 * \log h_2 = \log(h_1 h_2).$$

So $h = h_1 h_2$ and $(g_1 g_2)\psi = h_1 h_2 = g_1\psi g_2\psi$. This means that ψ is indeed an automorphism of G.

So we have proved that the automorphism group of G is isomorphic to the group of those automorphisms of the finite dimensional rational vector space L which map the subset $\log G$ onto itself. Thus we have a representation of the automorphism group of G as a group of linear transformations of a finite dimensional rational vector space, i.e., we have "linearized" the automorphism group. This information can be exploited more fully by combining it with the following. Let us term G a *lattice nilpotent group* (see [69]) if $\log G$ is a lattice in L, i.e., if $\log G$ is additively a subgroup of L. Thus if G happens to be a lattice nilpotent group the automorphism group of G is *isomorphic to the group of those automorphisms of L which map the lattice* $\log G$ *onto itself.* We shall make use of this later. For the moment we record only the following theorem of C. C. Moore [69], which can be proved quite easily, directly, by using the Baker-Campbell-Hausdorff formula, induction and Proposition 4.1.

Theorem 4.5. *Every finitely generated torsion-free nilpotent group G is contained in a lattice nilpotent group E as a subgroup of finite index.*

E is sometimes referred to as a *lattice nilpotent envelope* of G.

It should be noted at this point that we have almost established an isomorphism between finitely generated (in the obvious sense, as $\mathfrak{D}$-groups) nilpotent $\mathfrak{D}$-groups and finite dimensional nilpotent lie algebras over Q. Indeed we have

Theorem 4.6. *The categories of nilpotent $\mathfrak{D}$-groups and rational nilpotent lie algebras are equivalent.*

Theorem 4.6 is due, in the finitely generated case to A. I. Mal'cev [63]. If follows in general by making use of an unpublished generalization of Jennings theorem, which is due to M. Lazard, viz: Let G be any torsion-free nilpotent group, then QG is residually nilpotent. Here we shall indicate how Theorem 4.6 is proved in the case of finitely generated $\mathfrak{D}$-groups. Thus let G be such a $\mathfrak{D}$-group. It can be shown by means of Jennings theorem and Proposition 4.1 that QG is residually nilpotent. Moreover, it turns out that $L = \log G$ is itself a rational nilpotent lie algebra. Now consider the Baker-Campbell-Hausdorff group on L. The mapping

$$g \longrightarrow \log g$$

is an isomorphism from G to L^* since

$$g_1 g_2 \longrightarrow \log(g_1 g_2) = (\log g_1) * (\log g_2).$$

Thus we have a way of going back to G from L which does not involve the underlying associative algebra. This gives us part of the equivalence between the respective categories. It follows also, in much the same way as in our discussion of the automorphisms of finitely generated torsion-free nilpotent groups above, that if ϕ is a homomorphism of the nilpotent $\mathfrak{D}$-group G into the nilpotent $\mathfrak{D}$-group H, then ϕ induces a homomorphism of L into M, where L is the lie algebra of G and M is the lie algebra of H. Furthermore, if ϕ is any homomorphism of the rational nilpotent lie algebra L into the rational nilpotent lie algebra M, then ϕ defines a homomorphism of L^* into M^* in the obvious way. It is in this way that Theorem 4.6 is established.

We remark in closing that if G and H are finitely generated torsion-free nilpotent groups such that $m(G) = m(H)$ then G and H have the same lie algebras. This follows during the course of the proof (which we barely sketched above) of Theorem 4.6. We shall have occasion to use it later.

4.6. The object of the rest of Chapter 4 is to prove that the automorphism groups of certain finitely presented groups *and* (not necessarily associative) rings are again finitely presented. Such results are not universally valid, for J. Lewin [56] has constructed an uncomplicated finitely presented group whose automorphism group is not even finitely generated. Now finitely generated nilpotent groups are finitely presented. Hence every finite extension of a finitely generated nilpotent group is finitely presented. The first application of the methods we have introduced above will be to prove

Theorem 4.7. *The automorphism group of a finite extension of a finitely generated nilpotent group is finitely presented.*

There are two classes of groups to which Theorem 4.7 can be immediately applied. The first of these arise as the fundamental groups of compact flat manifolds. In fact such fundamental groups are finite extensions of free abelian groups of finite rank (see e.g., [86]). So Theorem 4.7 immediately gives us the

Corollary 4.71. *The automorphism group of the fundamental group of a compact flat manifold is finitely presented.*

The second class consists of the so-called *supersolvable* groups, i.e., groups which possess an invariant polycyclic series. Every supersolvable group is a finite extension of a finitely generated nilpotent group (see, e.g., [31], p. 4). So again Theorem 4.7 applies and we have

Corollary 4.72. *The automorphism group of a supersolvable group is finitely presented.*

Corollary 4.72 has been superseded by a very recent theorem of L. Auslander [3] which asserts that the automorphism group of every polycyclic group is finitely presented. His proof makes use of lie-group theoretic techniques. Roughly speaking the proof we have given here may be regarded as a good model of Auslander's proof of the polycyclic result.

Proceeding in a slightly different direction let us term a ring *Z-finite* if it is additively finitely generated (i.e., as a module over Z, the ring of integers, it is finitely generated). We emphasize that we use the term "ring" in a broad sense, i.e., a non-empty set with two binary operations $+$ and $\cdot$ which is an abelian group relative to $+$, with $+$ and $\cdot$ interrelated by the distributive laws (cf. e.g., R. D. Schafer [81], p. 2). It is easy to see that a Z-finite ring is finitely presented. Our second application is

Theorem 4.8. *The automorphism group of a Z-finite ring is finitely presented.*

Theorem 4.8 also has a number of applications. For example we have

Corollary 4.81. *The automorphism group of a finitely generated nilpotent lie ring is finitely presented.*

We recall that a lie ring L is *nilpotent* if its lower central series

$$L = L_1 \geq L_2 \geq \cdots$$

terminates in 0 after a finite number of steps. (The lower central series of L is defined inductively by putting L_{i+1} equal to the lie subring generated by all products ab ($a \in L_i$, $b \in L$).)

One can, of course, formulate an analogous result for associative rings.

4.7. Comments on the proofs. The proofs of both Theorem 4.7 and Theorem 4.8 center around the notion of "algebraic group." We recall some of the needed definitions here. Thus let F be a field and let V be a vector space of (finite) dimension n over F. Furthermore, let $e_1, e_2, \cdots, e_n$ be a fixed basis for V. If T is a linear transformation of V, then T corresponds to a unique matrix (t_{ij}) over F with respect to this fixed basis. Now let $\mathcal{P} = \{p_\lambda(x_{11}, \cdots, x_{nn}) | \lambda \in \Lambda\}$ be a family of polynomials in the n^2 variables $x_{11}, \cdots, x_{nn}$ over F. We term T a *zero* of $\mathcal{P}$ if $p_\lambda(t_{11}, \cdots, t_{nn}) = 0$ for every $\lambda \in \Lambda$. A group of linear transformations of V (i.e., a subgroup of $GL(V)$) is called an algebraic group if its elements comprise the complete set of zeros of some such family $\mathcal{P}$ of polynomials. We shall term an additive subgroup B of the vector space V a *lattice in* V if B is generated by some basis for V.

The relevance of these notions to finitely presented groups arises from the following theorem of A. Borel and Harish-Chandra [15]: *Let G be an algebraic group over the field Q of rational numbers and let B be a lattice in the underlying vector space V. Then the subgroup of G consisting of those linear transformations which map B onto B is finitely presented.*

The pertinence of this theorem is that the group of automorphisms of a finite dimensional algebra is an algebraic group. Both Theorem 4.7 and Theorem 4.8 will be reduced to the theorem of Borel and Harish-Chandra by making use of this fact. Indeed we obtain immediately from the Borel-Harish-Chandra theorem and the above discussion the result that the automorphism group of a lattice nilpotent group is finitely presented.

In the course of the proofs we shall prove some results which seem to be of independent interest. Before stating these results we recall that two groups (rings) are *commensurable* if they contain isomorphic normal subgroups (ideals) whose corresponding factor groups (rings) are finite. The automorphism groups of commensurable groups need not be commensurable; for example, the infinite cyclic group Z and the infinite dihedral group D are commensurable, but the automorphism group of D is infinite whereas that of Z is finite. However, denoting the automorphism group of a group G by $\operatorname{aut} G$ we shall prove, by residual techniques, the following result (G. Baumslag [13]):

Let G and H be commensurable, finitely generated nilpotent groups. Then $\operatorname{aut} G$ *and* $\operatorname{aut} H$ *are commensurable.*

The corresponding result for rings is rather easier to prove:

The automorphism groups of commensurable Z-finite rings are commensurable.

4.8. Let F be a torsion-free nilpotent group and let $m(F)$ be a (Mal′cev) completion of F. Recall that every automorphism α of F extends uniquely to an automorphism $m(\alpha)$ of $m(F)$. So, in the case of a torsion-free nilpotent group F, the mapping $\alpha \dashrightarrow m(\alpha)$ is an injection of $\operatorname{aut} F$ into $\operatorname{aut} m(F)$.

We say that a group X *acts* on a group G if it comes equipped with a homomorphism ϕ of X into $\operatorname{aut}(G)$. As usual X acts *faithfully* on G if ϕ is a monomorphism. If F is a subgroup of G then we define $\operatorname{stab}(F, X)$, the *stabilizer of F in X* by

$$\operatorname{stab}(F, X) = \{x \in X \mid Fx = F\},$$

where $Fx = \{fx \mid f \in F\}$; of course $fx = f(x\phi)$. It is clear that $\operatorname{stab}(F, X)$ is a subgroup of X. We shall often be concerned with a specific homomorphism of $\operatorname{stab}(F, X)$ into $\operatorname{aut}(F)$ which will be labeled res. Indeed if $y \in \operatorname{stab}(F, X)$, then res maps y into the automorphism y induces in F: $f(y\operatorname{res}) = fy$, $f \in F$.

We need a few elementary facts about res and $\operatorname{stab}(F, X)$.

Lemma 4.3. *Let the group X act on the finitely generated group G by* $\phi: X \dashrightarrow \operatorname{aut} G$, *and let F be a subgroup of finite index in G. Then* $\operatorname{stab}(F, X)$ *is of finite index in X.*

Proof. There are only finitely many subgroups of a given finite index in a finitely generated group. Let

$$F = F_1, F_2, \cdots, F_n \tag{1}$$

be the finitely many subgroups of the same finite index in G as F. Every element of $\operatorname{aut} G$ gives rise to a permutation of the set (1), providing us with a homomorphism ψ of $\operatorname{aut} G$ into a finite permutation group. Then $\ker \phi\psi$ is of finite index in X and is contained in $\operatorname{stab}(F, X)$. This proves Lemma 4.3.

Lemma 4.4. *Let G be a group and let F be a normal subgroup of finite index in*

G. Let res *be the restriction homomorphism of* stab$(F$, aut$G)$ *into* aut$F$, *and let* K *be the kernel of* res. *Then*

(a) *K is a finite extension of an abelian group;*

(b) *if G is a finite extension of a polycyclic group, then K is a finite extension of a finitely generated abelian group;*

(c) *if G is a finitely generated nilpotent group, then K is a finite group.*

Proof. (a) K maps F onto F. Thus the action of K on G induces an action of K on the finite group G/F. Hence there is a normal subgroup L of K of finite index which induces the identity automorphism in both G/F and F. But by a theorem of L. Kaloujnine [48], L is then abelian.

(b) Suppose now that N is a polycyclic normal subgroup of G of finite index. Then $N_1 = N \cap F$ is also a finitely generated polycyclic normal subgroup of G of finite index. By Lemma 4.3, stabN_1 is of finite index in aut$G$. Hence $S =$ stab$N_1 \cap$ stab$F$ is of finite index in stab$F$. So $K_1 = K \cap S$ is of finite index in K. K_1 induces in G/N_1 a finite group of automorphisms. Hence there is, as we saw above, an abelian normal subgroup L_1 of finite index in K_1 (and hence also in K) such that L_1 induces in both G/N_1 and N_1 the identity automorphism. So L_1 is abelian. Our object is to prove that L_1 is finitely generated.

To this end let $c_1, c_2, \cdots, c_k$ be a system of representatives of the cosets of N_1 in G. If $\alpha \in L_1$ then $c_i\alpha = a_{i,\alpha} c_i$ $(a_{i,\alpha} \in N_1)$. The elements $a_{i,\alpha} \in N_1$, $1 \le i \le k$, uniquely determine the automorphism α. Moreover if $\beta \in L_1$ is given by $c_i\beta = a_{i,\beta} c_i$, with $a_{i,\beta} \in N_1$, then $a_{i,\alpha} a_{i,\beta} = a_{i,\alpha\beta}$. It is then clear that the mapping of L_1 into N_1^k, the direct product of k copies of N_1, defined by

$$\alpha \longrightarrow (a_{1,\alpha}, \cdots, a_{k,\alpha}), \ (a \in L_1)$$

is a monomorphism. Since N_1^k is polycyclic, this implies that L_1 is finitely generated. Finally the fact that K has a finitely generated abelian subgroup L_1 of finite index implies that K has a finitely generated normal abelian subgroup $L_2 \le L_1$ still of finite index, and the proof of (b) is complete.

(c) It follows from (b) that we have only to prove that if L_1 is a (necessarily abelian) group of automorphisms of a finitely generated nilpotent group G which induces the identity automorphism in a normal subgroup F of finite index and the identity automorphism on the quotient group G/F, then L_1 is finite. Clearly, it is enough to prove that the $a_{i,\alpha}$ (in the notation above) are all of finite order. Observe that if G/F has order n, then $c_i^n \in F$. So if $\alpha \in L_1$, $c_i^n = c_i^n\alpha = (c_i\alpha)^n$. Now in a torsion-free nilpotent group, extraction of roots is unique. So $(c_i\alpha)c_i^{-1} \in \tau G$. In other words $a_{i,\alpha}$ is of finite order as required. This completes the proof of Lemma 4.4.

4.9. Ring-theoretical requirements. Let R be a ring. We denote the set of elements of R which are additively of finite order by τR. Clearly, τR is an ideal of R; τR is termed the *torsion-ideal* of R. R is said to be *torsion-free* if $\tau R = 0$. Clearly, $R/\tau R$ is torsion-free. R is *residually finite* if for each $r \in R$, $r \neq 0$, there is an ideal I of R with $r \notin I$ and R/I finite. The following simple result will be useful.

Lemma 4.5. *Let R be a Z-finite ring. Then R is residually finite.*

Proof. Define, for every integer n, $nR = \{nr | r \in R\}$. Then it is easy to see that nR is an ideal of R and, since R is additively a direct sum of a finite number of cyclic groups, that $\bigcap_{n=1}^{\infty} nR = 0$. Since R/nR is finite for every $n \geq 1$, this completes the proof of Lemma 4.5.

As with groups we denote the automorphism group of the ring R by aut R. So if R is a Z-finite ring, then aut R is residually finite. To see this we think of R additively, as a finitely generated abelian group. As such its (additive) automorphism group is residually finite since the abelian group R is finitely generated and residually finite (see [10]). So aut R is a subgroup of a residually finite group and is therefore residually finite.

4.10. The proof of Theorem 4.7: First reduction. Let G be a group with a finitely generated normal nilpotent subgroup F of finite index. The content of Theorem 4.7 is that aut G is finitely presented. The first step in the proof of Theorem 4.7 is to achieve the

First reduction. We may assume that F is torsion-free and that G splits over F.

In order to make this reduction, observe first that a finitely generated nilpotent group F has a torsion-free subgroup of finite index. Since F is of finite index in G it follows that there is a torsion-free normal subgroup of G of finite index in G contained in F. Therefore we may assume that F is torsion-free.

Let us now choose a system C of representatives $c_1, \cdots, c_k$ of the cosets of F in G. We denote the representative of the coset gF $(g \in G)$ by $\tilde{g}$. Thus $\tilde{g} = \tilde{h}$ if and only if g and h lie in the same coset. Every element g in G can then be written uniquely in the form $g = cf$ $(c \in C,\ f \in F)$. If $c, d \in C$, then we put $cd = \widetilde{cd}(c, d)$; so $(c, d) \in F$. Finally observe that each representative c_i gives rise to an automorphism γ_i of F via conjugation:

$$\gamma_i\colon f \to c_i^{-1} f c_i \quad (f \in F).$$

These definitions allow us to express the composition in G as follows:

$$c_i a \cdot c_j b = \widetilde{c_i c_j}(c_i, c_j)(a\gamma_j)b \quad (c_i, c_j \in C,\ a, b \in F).$$

Now let $\bar{F}$ be a (Mal'cev) completion of F. Every automorphism α of F extends uniquely to an automorphism $\bar{\alpha}$ of $\bar{F}$. Let $\bar{G}$ be the set of (formal) products ca $(c \in C,\ a \in \bar{F})$ and define a multiplication in $\bar{G}$ by

$$c_i a \cdot c_j b = \widetilde{c_i c_j}(c_i, c_j)(a\bar{\gamma}_j)b \quad (c_i, c_j \in C,\ a,\ b \in \bar{F}).$$

Then $\bar{G}$ is a group and G and $\bar{F}$ may be considered as subgroups of $\bar{G}$. Clearly, $\bar{F}$ is a normal subgroup of $\bar{G}$, $\bar{F} \cap G = F$, $\bar{F}G = \bar{G}$ and $\bar{G}/\bar{F}$ ($\cong G/F$) is finite; in fact, $c_1, \cdots, c_k$ may be used as coset representatives of $\bar{F}$ in $\bar{G}$. By Proposition 2.3, $\bar{G}$ splits over $\bar{F}$, so that we can find elements $a_1, \cdots, a_k \in \bar{F}$ such that $\{c_1a_1, \cdots, c_ka_k\} = E$ is a subgroup of $\bar{G}$.

Now it follows easily from Lemma 2.8 that the set of all elements of $\bar{F}$ with the property that some positive power lies in F is a group, hence a $\mathfrak{D}$-group, containing F. By the minimality of $\bar{F}$ we see that every element of $\bar{F}$ has some positive power lying in F. Choose an integer $l \geq 1$ such that $a_i^l \in F$ for $i = 1, \cdots, k$, and define $F_1 = gp(x \in \bar{F} | x^l \in F)$. It follows easily from the fact that F is a finitely generated subgroup of a torsion-free nilpotent group that F is of finite index in F_1. Moreover every automorphism α of F extends uniquely to an automorphism α_1 of F_1. For observe that if $x \in F_1$ and $\alpha \in \operatorname{aut} F$, then

$$(x\bar{\alpha})^l = x^l\bar{\alpha} = x^l\alpha \in F.$$

So $\bar{\alpha}$ induces in F_1 an automorphism α_1, say, the desired extension of α. Therefore, in particular, F_1 is normalized by E. We now define $G_1 = E \cdot F_1$; thus G_1 splits over F_1. To prove that $\operatorname{aut} G$ is finitely presented, it suffices to prove $\operatorname{aut} G_1$ is finitely presented (so that our first reduction is complete) for we shall now prove

Lemma 4.6. *$\operatorname{aut} G$ and $\operatorname{aut} G_1$ are commensurable.*

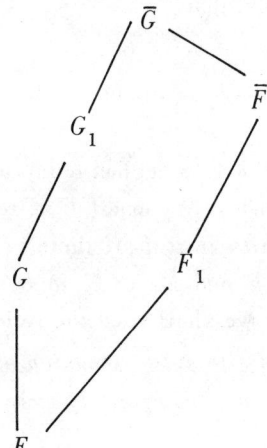

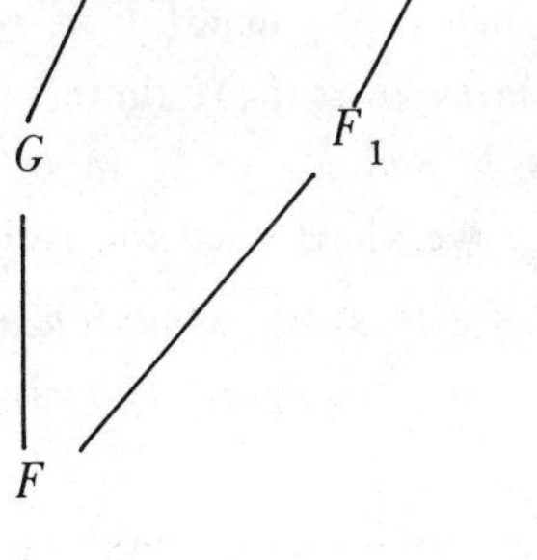

Note first that every element g in G_1 can be written uniquely in the form

$$g = cf \ (c \in C,\ f \in F_1).$$

Now consider the stabilizer S of F in aut G. By Lemma 4.3, S is of finite index in aut G. Moreover, every automorphism σ in S extends uniquely to an automorphism σ_1 of G_1. To see this notice that σ induces in F an automorphism λ. This automorphism λ extends uniquely to an automorphism λ_1 of F_1. Define a mapping σ_1 of G_1 into G_1 by

$$\sigma_1 : cf \to (c\sigma)(f\lambda_1) \ (c \in C, f \in F_1).$$

It can be verified that σ_1 is an automorphism of G_1 by a direct computation.

Consider now the full automorphism group aut G_1 of G_1. Let U be the stabilizer of F_1 in aut G_1 and V the stabilizer of G in aut G_1. Since both F_1 and G are of finite index in G_1, U and V are of finite index in aut G_1 (Lemma 4.3). So $W = U \cap V$ is of finite index in aut G_1. Let ρ be the restriction homomorphism of W into aut G. We claim that ρ is a monomorphism. For if $\alpha \in W$ and $\alpha\rho = 1$, then α induces the identity automorphism on F. Since α maps F_1 onto F_1 it follows from the fact that extraction of roots is unique in F_1 that α induces the identity on F_1. So α must be the identity automorphism on $GF_1 = G_1$ since α induces the identity on G itself.

Finally we show that $W\rho \geq S$, so that $W\rho$ is of finite index in aut G. For if $\sigma \in S$, then the associated automorphism σ_1 of G_1 maps F_1 onto F_1 and G onto G, that is $\sigma_1 \in W$; and clearly $\sigma_1\rho = \sigma$. It follows immediately that aut G_1 and aut G are commensurable, since W is of finite index in aut G_1 and $W\rho$ is of finite index in aut G. This completes the proof of the lemma.

We observe now that if Z is central in G then Z is central in G_1. It follows almost immediately by induction that

Lemma 4.7. *If G is nilpotent so is G_1.*

This completes the first stage of the proof. In particular we have made our first reduction.

4.11. The proof of Theorem 4.7: Second reduction. Let G be a group with a normal subgroup F of finite index. As usual E is termed a *complement of* F if $F \cap E = 1$ and $FE = G$. G *splits* over F if there is at least one complement of F in G. We term two complements E and E_1 of F in G *equivalent* if there is an element $a \in F$ such that $a^{-1}Ea = E_1$. We shall need the following

Lemma 4.8. *Let G be a finite split extension of a finitely generated torsion-free nilpotent group F. Then there are only finitely many inequivalent complements of F in G.*

Proof. We proceed by induction on the class of F. If F is abelian then the number of such inequivalent complements is simply the order of the first cohomology group $H^1(E, F)$ of some complement E of F where F is thought of as an E-module in the obvious way. But $H^1(E, F)$ is finite since F is a finitely generated torsion-free abelian group and E is finite.

Suppose now that F is not abelian. Let Z be the center of F. Then Z is a r normal subgroup of G. Let furthermore $\{E_\lambda | \lambda \in \Lambda\}$ be the family of all complements of F in G. Now G/Z is a splitting extension of F/Z. Moreover, F/Z is torsion-free. Therefore, there are only finitely many inequivalent complements. This means that we can find $E_1Z/Z, E_2Z/Z, \cdots, E_nZ/Z$ such that every $E_\lambda Z/Z$ is equivalent to one of these complements. But observe now that if $\lambda \in \Lambda$ and $E_\lambda Z/Z$ is conjugate to E_iZ/Z, then we can find $f \in F$ such that $f^{-1}(E_\lambda Z)f = E_iZ$. Consider $H = E_iZ$. H is a splitting extension of the torsion-free abelian group Z by the subgroup E_i. Consequently, there are only finitely many inequivalent complements of Z in H, say

$$E = E_{i,1}, E_{i,2}, \cdots, E_{i,m(i)}.$$

Therefore, noting that $f^{-1}E_\lambda f$ is a complement of Z, we can find $z \in Z$ such that $z^{-1}f^{-1}E_\lambda fz = E_{i,\lambda(i)}$ for some $\lambda(i)$, $1 \leq \lambda(i) \leq m(i)$. Thus we have shown that every complement of F in G is equivalent to one of the finitely many complements

$$E_{1,1}, \cdots, E_{1,m(1)}, E_{2,1}, \cdots, E_{2,m(2)}, \cdots, E_{n,1}, \cdots, E_{n,m(n)}.$$

Corollary 4.81. *Let G be a finite split extension of a finitely generated torsion-free nilpotent group F. Then there is a subgroup U of finite index in* aut G *which stabilizes F and leaves the equivalence classes of complements of F in G invariant.*

Proof. By Lemma 4.3, the stabilizer T of F in aut G is of finite index in aut G. If $\tau \in T$, then $F\tau = F$, so if E is any complement of F in G, then clearly $E\tau$ is again a complement of F in G. Furthermore, if $f \in F$, then

$$(f^{-1}Ef)\tau = (f\tau)^{-1}E\tau(f\tau).$$

Since $f\tau \in F$, it follows that τ permutes the equivalence classes of complements of F in G. By Lemma 4.8, there are only finitely many such equivalence classes. So there is a subgroup U of finite index in T, namely the kernel of this homomorphism of T into the permutation group on the equivalence classes of complements, such that that if $\mu \in U$ then μ maps every equivalence class of complements into itself. This completes the proof.

Let us assume for the rest of Chapter 4 that G is a split extension of a finitely generated torsion-free nilpotent subgroup F by a finite group E.

We shall make use of the following notation. If $g \in G$, then $\hat{g} \in$ aut F is given by $\hat{g}: x \to g^{-1}xg$ $(x \in F)$. Of course the mapping $g \to \hat{g}$ is a homomorphism of G into aut F. We denote by $\hat{E}$, $\hat{F}$ the images of E and F respectively under this homomorphism. Observe that $\hat{F}$ is the group of inner automorphisms of F. Hence $\hat{F}$ is normal in aut F. We denote by ρ the restriction homomorphism of the subgroup U of

Corollary 4.81 into aut F. Finally let C be the centralizer of $\hat{E}$ in aut F and let N be normalizer of $\hat{E}$ in aut F.

We may formulate the next step in the proof of Theorem 4.7 in these terms as

Second reduction. It is enough to prove that C is finitely presented.

We shall actually obtain a more precise connection between C and aut G, most of which is embodied in the following

Lemma 4.9. *$U\rho$ is of finite index in $\hat{F}N$, $\hat{F}C$ is of finite index in $\hat{F}N$ and the kernel of ρ is a finite extension of a finitely generated abelian group.*

Proof. An inspection of the definition of U shows that $\hat{F} \leq U\rho$. To see that $U\rho$ contains C let $\alpha \in C$. Define a mapping α^* of G into G by

$$(fe)\alpha^* = f\alpha \cdot e \quad (f \in F,\ e \in E).$$

It follows readily that α^* is an automorphism of G. Therefore, by its very definition, $\alpha^* \in U$. Obviously then $\alpha^*\rho = \alpha$. So $C \leq U\rho$. Diagramatically then we have the following situation:

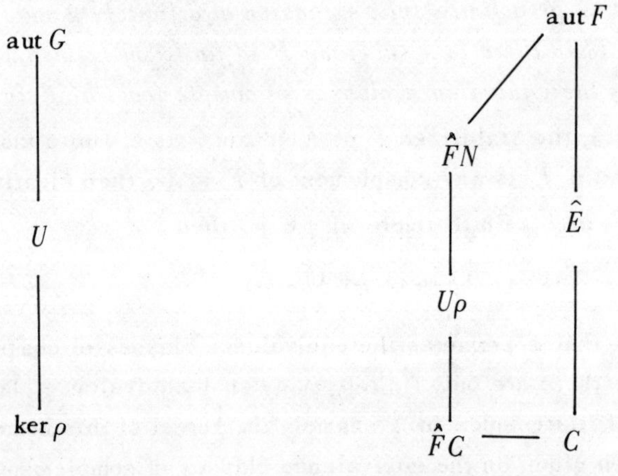

On the other hand suppose $\mu \in U$. Then there is an $f \in F$ such that $E\mu = f^{-1}Ef = E\hat{f}$. Therefore $\hat{f}(\mu\rho)^{-1}\hat{E}\mu\rho\hat{f}^{-1} = \hat{E}$. So $\mu\rho\hat{f}^{-1} \in N$ or, in other words, $U\rho \leq \hat{F}N$.

Now since E is finite, it follows that C is of finite index in N, so that $\hat{F}C$ is of finite index in $\hat{F}N$. Finally the fact that $\ker\rho$ is a finite extension of a finitely generated abelian group follows at once from Lemma 4.4. This completes the proof.

We note that Lemma 4.9 does achieve our second reduction. For suppose we can show that C is finitely presented; we claim that it follows that aut G is finitely presented. For $\hat{F} \cap C$ is finitely generated, being a subgroup of the finitely generated nilpptent group $\hat{F}$; therefore $C/\hat{F} \cap C \cong \hat{F}C/\hat{F}$ is finitely presented. From this we see that $\hat{F}C$, being an extension of the finitely presented group $\hat{F}$ by $\hat{F}C/\hat{F}$, is finitely presented. As $U\rho$ is of finite index in $\hat{F}C$, $U\rho$ is finitely presented. Therefore U is finitely presented because U is an extension of the finitely presented group $\ker\rho$ by $U\rho$. Finally we see that $\operatorname{aut} G$ too is finitely presented because it contains U as a subgroup of finite index.

4.12. Completion of the proof of Theorem 4.7. Let F be a finitely generated, torsion-free nilpotent group. Let L be the lie algebra of F. As we remarked before, $\operatorname{aut} F$ is isomorphic to the subgroup of $\operatorname{aut} L$ which maps $\log F$ onto $\log F$. Let now X be a finite subgroup of $\operatorname{aut} F$ and let $\tilde{X}$ be the image of X under the isomorphism from $\operatorname{aut} F$ into $\operatorname{aut} L$ mentioned above. Since $\operatorname{aut} L$ is an algebraic group so is $\tilde{C}$, the centralizer of $\tilde{X}$ in $\operatorname{aut} L$ (see [15]). Now choose, via Theorem 4.5, a lattice nilpotent envelope E of F. As we remarked at the end of the proof of Theorem 4.6, L is also the lie algebra of E. Thus $\log E$ is a lattice in L. Now by the theorem of Borel and Harish-Chandra the group C^* of those elements of $\tilde{C}$ which map the lattice $\log E$ onto itself is finitely presented. Suppose now we can show that $\operatorname{aut} E$ and $\operatorname{aut} F$ are commensurable. Then the centralizer C of X in $\operatorname{aut} F$ will be commensurable with C^* and consequently finitely presented. According to our second reduction, this will complete the proof of Theorem 4.7.

So Theorem 4.7 is established save for the proof of the following result proved in [13], which is of independent interest.

Theorem 4.9. *Let G and H be finitely generated nilpotent commensurable groups. Then* $\operatorname{aut} G$ *and* $\operatorname{aut} H$ *are commensurable.*

The proof of Theorem 4.9 will be carried out by proving a succession of lemmas.

Lemma 4.10. *Let G be a finitely generated torsion-free nilpotent group and let F be a subgroup of finite index. Then* $\operatorname{aut} G$ *and* $\operatorname{aut} F$ *are commensurable.*

Proof. Suppose F is of index k in G. Let $\bar{G}$ be a Mal'cev completion of G and let $F_1 = gp(x \in \bar{G} \mid x^k \in F)$. Then F_1 contains G as a subgroup of finite index and, as we observed immediately before Lemma 4.6, every automorphism α of F extends uniquely to an automorphism α_1 of F_1. Hence if S is the stabilizer of F in $\operatorname{aut} F_1$ and ρ is the restriction homomorphism of S into $\operatorname{aut} F$, then

$$S\rho = \operatorname{aut} F.$$

But there is a subgroup T of finite index in S such that T stabilizes G. Therefore, $T\rho$ is of finite index in $\operatorname{aut} F$. It follows that if U is the stabilizer of F in $\operatorname{aut} G$ and if σ is the restriction homomorphism of U into $\operatorname{aut} F$ then $U\sigma \geq T\rho$. So $U\sigma$ is of finite index in $\operatorname{aut} F$. Now U is of finite index in $\operatorname{aut} G$. Furthermore, σ is a monomorphism since F is of finite index in the torsion-free nilpotent group G. This means $\operatorname{aut} G$

and aut F are commensurable.

Actually Lemma 4.10 is really all we need to finish off the proof of Theorem 4.7. It is crucial in the proof of Theorem 4.9.

The next result we need is

Lemma 4.11. *Let P be the direct product of a finitely generated torsion-free nilpotent group A and a finite group B. Let, further, A_1 be a normal subgroup of A of finite index and let G be a subgroup of P containing A_1. Then* aut G *and* aut A_1 *are commensurable.*

Proof. Now there is a subgroup P_1 of finite index in aut P which maps A_1 onto A_1, A onto A and G onto G. Every element of P_1 therefore induces, in particular, a permutation of the finitely many cosets of A in P. So there is a subgroup P_2 of finite index in P_1 such that P_2 leaves every coset of A_1 in P invariant. Now the elements of P_2 give rise to automorphisms of A_1 by restriction. We denote the corresponding homomorphism of P_2 into aut A_1 by ρ. The kernel K of ρ consists of those automorphisms of P which induce simultaneously the identity automorphism on the factor group P/A_1 and on A_1 itself. It is not difficult to see that $K = 1$ since A_1 is torsion-free. This means that ρ is a monomorphism.

We shall reformulate our maneuvers as

(a) There is a subgroup X $(= P_2)$ of aut P of finite index such that X maps A_1 onto A_1, A onto A and G onto G. Moreover the restriction homomorphism ρ of X to A_1 is a monomorphism.

We now make use of Lemma 4.10 to find a subgroup Z_1 of aut A_1 of finite index such that every element ζ of Z_1 can be extended uniquely to an automorphism ζ^* of A. Let ι_1 be the mapping defined by $\iota_1\colon \zeta \to \zeta^*$. Then by the uniqueness of ζ^* it follows that ι_1 is a monomorphism of Z_1 into aut A. Since $P = A \times B$ every automorphism of A extends uniquely to an automorphism of P by stipulating that it act as the identity automorphism on B. The resultant mapping ϵ is clearly a monomorphism of aut A into aut P. Put $\iota = \iota_1 \epsilon$. Then ι is a monomorphism of Z_1 into aut P.

Next we choose a subgroup Y of finite index in aut G in much the same way as we chose X above. To be precise we choose Y so that

(b) Y maps A_1 onto A_1, the restriction homomorphism ρ_1 of Y into A_1 is a monomorphism, $Y\rho_1 \subseteq Z_1$, $Y\rho_1\iota \subseteq X$ and Y is of finite index in aut G.

Finally we choose $Z \subseteq$ aut A_1 so that

(c) Z is of finite index in aut A_1, $Z\iota \subseteq X$ and $Z\iota\rho_2 \subseteq Y$ where ρ_2 is the restriction homomorphism of X to G (since X maps G onto G, ρ_2 makes sense).

Notice now that as ρ itself is a monomorphism, ρ_2 is also a monomorphism on X and so ρ_2 is a monomorphism on Z. It is clear that if $\alpha \in Z$, then

$\alpha\iota\rho_2\rho_1 = \alpha\iota\rho = \alpha$. Therefore by (a), (b) and (c) we have $Y\rho_1 \supseteq (Z\iota\rho_2)\rho_1 = Z$. So $Y\rho_1$ is of finite index in aut A_1. But ρ_1 is a monomorphism. Hence, remembering that Y is of finite index in aut G we have shown that aut A_1 and aut G are commensurable. This completes the proof of the lemma.

The proof of Theorem 4.9 now follows easily. For let G be any finitely generated nilpotent group. Let T be the torsion-subgroup of G. Then T is finite. Let, further, U be a normal torsion-free subgroup of finite index in G. Then $U \cap T = 1$; so G can be embedded in $P = A \times B$ where $A = G/T$ and $B = G/U$. Of course, A is torsion-free and B is finite. Moreover, the image of G in P certainly contains the image A_1 of U which is of finite index in A. So, by Lemma 4.11, aut A_1 and aut G are commensurable. Hence, by Lemma 4.10, aut G is commensurable with aut C for every torsion-free subgroup C of finite index in G. Clearly then if G and H are commensurable finitely generated nilpotent groups, aut G and aut H are commensurable. This completes the proof of Theorem 4.9.

The proof of Theorem 4.8 is similar to that of Theorem 4.7 and will not be given here.

Chapter 5. Miscellaneous topics

In this chapter we shall record, and briefly discuss, some theorems related to the study of finitely generated nilpotent groups.

5.1. Frattini proved that the intersection ΦG of the maximal subgroups of a finite group G is nilpotent (ΦG is called the *Frattini subgroup* of G). Here we shall sketch the proof of the recent generalisation of Frattini's theorem due to V. P. Platonov [77] (cf. also Wehrfritz [85]).

Theorem 5.1. *The Frattini subgroup of a finitely generated group of matrices over a field is nilpotent.*

(We use the term field for a commutative division ring.)

We begin with a sketch of the proof of the following theorem of B. H. Neumann [73] from which we will eventually deduce Theorem 5.1.

Theorem 5.2. *If a finitely generated commutative ring R is a field then it is finite.*

To see how Theorem 5.2 is proved consider a basis B of transcendental elements over the prime field P of R. Since R is a finitely generated ring R is a finite extension of the subfield Q generated by B. Now it is not difficult to prove that B is empty because in any polynomial ring over a field there exists infinitely many primes. Finally it follows similarly from the corresponding fact about the ring of rational integers that P is a finite field.

(For a slightly different, slightly slicker (but still easy) complete proof we refer the reader to B. H. Neumann's paper [73].)

Theorem 5.2 plays an essential role in the proof of

Theorem 5.3. *A finitely generated commutative ring R is residually finite.*

We postpone the proof of Theorem 5.3 in order to record (and prove) some interesting applications. The first of these is the proof of a special case of a theorem of P. Hall [34]:

Theorem 5.4. *A finitely generated metabelian group G is residually finite.*

Theorem 5.4 will follow from the following

Proposition. *A finitely generated module M over a commutative ring R whose quotients are all residually finite is residually finite.*

Proof. Let $a \in M$ ($a \neq 0$). Suppose N is a submodule of M which is maximal subject to $a \notin N$. We shall prove M/N is finite. Thus it is enough to consider the case where M is such that every nonzero submodule of M contains a. Now M is finitely generated, say by $a_1, \cdots, a_k$. Let M_i be the submodule of M generated by a_i. Now $a \in M_i$ $(i = 1, \cdots, k)$. But $M_i \cong R/A_i$ where A_i is the annihilator of a_i. Hence M_i is residually finite. So we can find a submodule L_i of M_i with $a \notin L_i$ and M_i/L_i finite. But every nonzero submodule of M contains a. So $L_i = 0$ and M_i is finite. Hence $M = M_1 + \cdots + M_k$ is finite, as desired.

Theorem 5.4 follows easily. For we may think of G' as an R $(= Z(G/G'))$-module (where here G' denotes the derived group of G). Now if $G = gp(x_1, \cdots, x_q)$ then G' is the normal subgroup generated by the commutators $[x_i, x_j]$ $(1 \leq i < j \leq q)$. So G' is a finitely generated R-module. By Theorem 5.3 and the Proposition, G' is residually finite as an R-module. This means that G is residually-finite-by-finitely generated abelian. Hence G is residually polycyclic and therefore residually finite. This completes the proof of Theorem 5.4.

P. Hall has proved somewhat more than Theorem 5.4 (see [34]). One (by no means easy) consequence of Theorem 5.4 is the following beautiful theorem of P. Hall [35].

Theorem 5.5. *The Frattini subgroup of a finitely generated metabelian group is nilpotent.*

We come now to the proof of Theorem 5.3. Actually, I have not been able to locate a complete proof of Theorem 5.3 in the literature. The proof I shall give here is based on Theorem 5.2 and an argument of N. Divinsky [20]. In the case of an integral domain it is due to A. I. Mal'cev [65].

Proof of Theorem 5.3. Let $a \in R$ $(a \neq 0)$ and let I be a maximal ideal of R which does not contain a. It is enough to prove that R/I is finite. We may therefore assume $I = 0$ and hence that R is a finitely generated, commutative ring with an element $a \neq 0$ such that every nonzero ideal of R contains a. Let M be the minimal ideal of R containing a. Then M is actually the unique minimal nonzero ideal of R. Let A be the annihilator of M. Then either M is a faithful irreducible R/A-module or else $A = R$. If $A \neq R$, R/A is a field (by Schur's lemma). So R/A is finite by Theorem 5.2. So in both cases R/A is finite. Therefore A is, like R, a finitely generated commutative ring.

We shall prove that A is nilpotent. Since A is finitely generated it is enough to prove that every element of A is nilpotent. Suppose, if possible, that $u \in A$ is not nilpotent. Then $u^i \neq 0$ for every $i = 1, 2, \cdots$. So $u^iA \neq 0$. But u^iA is an ideal of R and so by the unique minimality of M, $u^iA \geq M$ for every $i = 1, 2, \cdots$. Thus we can find elements $y_1, y_2, \cdots$ in A such that $u^iy_i = a$ $(i = 1, 2, \cdots)$. Of course $u^{i+1}y_i = ua = 0$. So y_i is in the annihilator A_{i+1} of u^{i+1} although $y_i \notin A_i$, the annihilator of y_i $(i = 1, 2, \cdots)$. So $A_1 \subset A_2 \subset \cdots$ is an infinite properly ascending chain of ideals of R. But R is a finitely generated commutative ring and hence noetherian.

This contradiction establishes the nilpotence of u and therefore also that of A. So A is additively a finitely generated abelian group. Consequently this is true also of R. Therefore R is a Z-finite ring in the sense of Chapter 4 and hence residually finite. So R is finite as claimed.

It follows from Theorem 5.3 that the Jacobson radical of a finitely generated commutative ring is residually nilpotent (Krull [51]). This theorem should be compared with the corresponding theorem of Philip Hall in [35] on the nature of the Jacobson radical of a quotient ring of the group ring of a finitely generated nilpotent group.

Theorem 5.3 allows us to prove a slight generalisation of the following theorem of A. I. Mal'cev [65].

Theorem 5.6. *A finitely generated group of matrices over a commutative ring is residually finite.*

Proof. Let G be a finitely generated group of matrices, of degree n, over a commutative ring, say

$$G = gp((a_{ij}^{(1)}), \cdots, (a_{ij}^{(n)})) \ (n < \infty).$$

Then G is the semigroup generated by the matrices

$$(a_{ij}^{(1)}), \cdots, (a_{ij}^{(n)}), (b_{ij}^{(1)}), \cdots, (b_{ij}^{(n)}) \quad (1)$$

where here

$$(b_{ij}^{(k)}) = (a_{ij}^{(k)})^{-1} \ (\text{for } k = 1, \cdots, n).$$

Let R be the subring generated by the coefficients of the matrices listed in (1). If ϕ is any homomorphism of R onto a ring S then ϕ induces a homomorphism ϕ_* of the ring of all $d \times d$ matrices over R onto the corresponding ring of matrices over S. Now let $(g_{ij}) \in G$. If (g_{ij}) is not the identity matrix it follows easily from the residual finiteness of R that we can choose a homomorphism ϕ of R onto a finite ring S such that $(g_{ij})\phi_* \neq 1$.

Since the totality of all $d \times d$ matrices over S is finite, $G\phi_*$ is finite. This completes the proof of Theorem 5.6.

For finitely generated matrix groups over a field the image rings S that arise can be controlled sufficiently so as to yield a slightly stronger result than Theorem 5.6. Theorem 5.1 can be deduced from this stronger form of Theorem 5.6 on combining it with known results on nilpotent subgroups of matrix groups over finite fields (for further details see [77]).

5.2. Suppose that G is a finitely generated torsion-free nilpotent group. Then we can find a central series $G = G_0 > G_1 > \cdots > G_q = 1$ in which successive factors are infinite cyclic, say $G_i = gp(a_{i+1}, G_{i+1})$ $(i = 0, 1, \cdots, q-1)$. It follows that if $g \in G$ then g can be written uniquely in the form

$$g = a_1^{\gamma_1} a_2^{\gamma_2} \cdots a_q^{\gamma_q} \ (\gamma_i \in Z).$$

Thus the mapping $\phi: g \longrightarrow (\gamma_1, \cdots, \gamma_q)$ is a $1-1$ mapping of the set G into the direct product of q copies of Z. Now A. I. Mal'cev [63] has pointed out that there exist polynomials $f_1, f_2, \cdots, f_q$ in $2q$ variables over Z, such that if

$$g\phi = (\gamma_1, \cdots, \gamma_q) \text{ and } h\phi = (\delta_1, \cdots, \delta_q)$$

then

$$(gh)\phi = (f_1(\gamma_1, \cdots, \gamma_q, \delta_1, \cdots, \delta_q), \cdots, f_q(\gamma_1, \cdots, \gamma_q, \delta_1, \cdots, \delta_q)).$$

Similarly inversion in G can be described by a polynomial in q-variables over Z. Hence if we define a composition in R^q by

$$(\gamma_1, \cdots, \gamma_q) \cdot (\delta_1, \cdots, \delta_q) = (f_1(\gamma_1, \cdots, \gamma_q, \delta_1, \cdots, \delta_q), \\ \cdots, f_q(\gamma_1, \cdots, \gamma_q, \delta_1, \cdots, \delta_q))$$

and similarly for inversion, this turns R^q into a lie group. It follows that G is embedded as a discrete subgroup of this lie group in such a way that the quotient space is compact. So we have formulated most of another theorem of A. I. Mal'cev [63], (which we have actually proved already in Chapter 4 via Jenning's procedure) namely

Theorem 5.7. *Every finitely generated torsion-free nilpotent group can be embedded as a uniform subgroup of a connected, simply connected real lie group.*

The classical connection between lie groups and lie algebras can now be utilized to provide us with information about G (see, e.g., [86] for a discussion of lie groups and lie algebras). Our approach has been motivated by these classical facts.

Nilpotent groups play an important role in the structure of polycyclic groups. The main result here is the following theorem of A. I. Mal'cev [62] which was motivated by a theorem of Kolchen on matrix groups.

Theorem 5.8. *Every polycyclic group contains a subgroup of finite index whose derived group is nilpotent.*

The distinction between Mal'cev's result and Kolchen's disappeared when L. Auslander [2] (see R. Swan [83] for a different proof) settled a problem raised by P. Hall in the 1950's by proving

Theorem 5.9. *Every polycyclic group has a faithful representation in a group of matrices over the rational integers.*

Keeping our attention on matrix groups we have the following theorem.

Theorem 5.10. *Every solvable group of matrices over the ring of integers of an algebraic number field is polycyclic.*

Theorem 5.10 yields two further theorems of A. I. Mal′cev [62] namely:

Theorem 5.11. *The solvable subgroups of the group of automorphisms of a polycyclic group are polycyclic.*

Theorem 5.12. *A solvable group is polycyclic if and only if all its abelian subgroups are finitely generated.*

We turn our attention now to a rather different notion due to J. Milnor [67] which essentially singles out the nilpotent groups from other solvable groups. Milnor's notion is that of growth function. In order to explain let G be a group with a given finite set $X = \{x_1, \cdots, x_q\}$ of generators. For each positive integer n let $g(n)$ denote the number of elements of G that can be expressed as words of length at most n. As it stands $g(n)$ is a function of the positive integer n. The function g is called a *growth function* for G. We term g of polynomial type of degree $\leq e$ if $g(n) \leq cn^e$ where c is some constant. If $g(n) \geq uv^n$ where both u and v are constants, $v > 1$, then g is said to be of exponential type. It turns out that if any growth function of G is of a given type, then they all are, i.e., the choice of the finite system of generators of G does not affect the type of the associated growth function. The relevance of these notions to us lies in the following result due in part to J. Milnor and in part to J. A. Wolf (see [87]).

Theorem 5.13. *A finitely generated solvable group is either of exponential type or of polynomial type. It is of polynomial type if and only if it is a finite extension of a nilpotent group.*

The relevance of this notion of a growth function to algebra is not clear, but it is still quite fascinating. It arose in connection with the study of curvature of certain Riemannian manifolds (see J. Milnor [68] and J. A. Wolf [87]).

BIBLIOGRAPHY

[1] L. Auslander, *Bieberbach's theorems on space groups and discrete uniform subgroups of Lie groups*, Ann. of Math. (2) 71 (1960), 579–590. MR 22 #12161.

[2] ———, *On a problem of Philip Hall*, Ann. of Math. (2) 86 (1967), 112–116. MR 36 #1540.

[3] ———, *The automorphism group of a polycyclic group*, Ann. of Math. (2) 89 (1969), 314–322.

[4] L. Auslander and G. Baumslag, *Automorphism groups of finitely generated nilpotent groups*, Bull. Amer. Math. Soc. 73 (1967), 716–717. MR 36 #259.

[5] R. Baer, *Endlicheitskriterien für Kommutatorgruppen*, Math. Ann. **124** (1952), 161–177. MR 13, 622.

[6] ———, *Engelsche elemente Noetherscher Gruppen*, Math. Ann. **133** (1957), 256–270. MR 19, 248.

[7] G. Baumslag, *Some aspects of groups with unique roots*, Acta Math. **104** (1960), 217–303. MR 23 #A191.

[8] ———, *A generalization of a theorem of Mal'cev*, Arch. Math. 12 (1961), 405–408. MR 26 #205.

[9] ———, *On the residual finiteness of generalized free products of nilpotent groups*, Trans. Amer. Math. Soc. 106 (1963), 193–209. MR 26 #2489.

[10] ———, *Automorphism groups of residually finite groups*, J. London Math. Soc. 38 (1963), 117–118. MR 26 #3793.

[11] ———, *Groups with the same lower central sequence as a relatively free group. I. The groups*, Trans. Amer. Math. Soc. 129 (1967), 308–321. MR 36 #248.

[12] ———, *Groups with the same lower central sequence as a relatively free group. II. Properties*, Trans. Amer. Math. Soc. 142 (1969), 507–538. MR 39 #6959.

[13] ———, *Automorphism groups of nilpotent groups*, American J. Math 91 (1969), 1003–1011.

[14] N. Blackburn, *Conjugacy in nilpotent groups*, Proc. Amer. Math. Soc. **16** (1965), 143–148. MR 30 #3140.

[15] A. Borel, *Arithmetic properties of linear algebraic groups*, Proc. Internat. Congress Math. (Stockholm, 1962) Inst. Mittag-Leffler, Djursholm, 1963, pp. 10–22. MR 31 #177.

[16] A. Borel and J-P. Serre, *Théorèmes de finitude en cohomologie galoisienne*, Comment. Math. Helv. **39** (1964), 111–164. MR 31 #5870.

[17] J. F. Bowers, *On composition series of polycyclic groups*, J. London Math. Soc. **35** (1960), 433–444. MR 23 #A1710.

[18] C. Chevalley, *Theory of Lie groups*. Vol. 1, Princeton Math. Series, vol. 8, Princeton Univ. Press, Princeton, N.J., 1946. MR **7**, 412.

[19] M. Dehn, *Über unendliche diskontinuierliche Gruppen*, Math. Ann. **71** (1911), 116–144.

[20] N. Divinsky, *Commutative subdirectly irreducible rings*, Proc. Amer. Math. Soc. **8** (1957), 642–648. MR **19**, 245.

[21] V. H. Dyson, *The word problem and residually finite groups*, Notices Amer. Math. Soc. **11** (1964), 743. Abstract #616–7.

[22] W. Gaschütz, *Kohomologische Trivialitäten und äussere Automorphismen von p-Gruppen*, Math. Z. **88** (1965), 432–433. MR **33** #4137.

[23] E. S. Golod, *On nil-algebras and finitely approximable p-groups*, Izv. Akad. Nauk SSSR Ser. Mat. 28 (1964), 273–276; English transl., Amer. Math. Soc. Transl. (2) **48** (1965), 103–106. MR **28** #5082.

[24] K. W. Gruenberg, *Two theorems on Engel groups*, Proc. Cambridge Philos. Soc. **49** (1953), 377–380. MR **14**, 1060.

[25] ———, *Residual properties of infinite soluble groups*, Proc. London Math. Soc. (3) 7 (1957), 29–62. MR **19**, 386.

[26] ———, *The upper central series in soluble groups*, Illinois J. Math. **5** (1961), 436–466. MR 25 #122.

[27] M. Hall, *The theory of groups*, Macimillan, New York, 1959. MR 21 #1996.

[28] ———, *Subgroups of finite index in free groups*, Canad. J. Math. **1** (1949), 187–190. MR **10**, 506.

[29] P. Hall, *A contribution to the theory of groups of prime-power order*, Proc. London Math. Soc. (2) 36 (1933), 29–95.

[30] ———, *Finiteness conditions for soluble groups*, Proc. London Math. Soc. (3) **4** (1954), 419–436. MR **17**, 344.

[31] ———, *Nilpotent groups*, Canad. Math. Congress, Summer Seminar, University of Alberta, 1957.

[32] ———, *Some word-problems*, J. London Math. Soc. 33 (1958), 482–496. MR 21 #1331.

[33] ———, *Finite-by-nilpotent groups*, Proc. Cambridge Philos. Soc. 52 (1956), 611–616. MR **18**, 190.

[34] ———, *On the finiteness of certain soluble groups*, Proc. London Math. Soc. (3) **9** (1959), 595–622. MR **22** #1618.

[35] ———, *The Frattini subgroups of finitely generated groups*, Proc. London Math. Soc. (3) **11** (1961), 327–352. MR 23 #A1718.

[36] H. Heineken, *Engelsche Elemente der Länge drei*, Illinois J. Math. **5** (1961), 681–707. MR 24 #A1319.

[37] G. Higman, *The units of group-rings*, Proc. London Math. Soc. (2) 46 (1940), 231–248. MR **2**, 5.

[38] ———, *A remark on finitely generated nilpotent groups*, Proc. Amer. Math. Soc. 6 (1955), 284–285. MR **16**, 996.

[39] ———, *Groups and rings having automorphisms without non-trivial fixed elements*, J. London Math. Soc. **32** (1957), 321–334. MR **19**, 633.

[40] ———, *Lie ring methods in the theory of finite nilpotent groups*, Proc. Internat. Congress Math. (Edinburgh, 1958) Cambridge Univ. Press, New York, 1960, pp. 307–312. MR 22 #6845.

[41] ———, Unpublished, 1969.

[42] K. A. Hirsch, *On infinite soluble groups*. I, Proc. London Math. Soc. (2) **44** (1938), 53–60.

[43] ———, *On infinite soluble groups*. II, Proc. London Math. Soc. (2) **44** (1938), 336–344.

[44] ———, *On infinite soluble groups*. III, Proc. London Math. Soc. (2) **49** (1946), 184–194. MR **8**, 132.

[45] N. Jacobson, *Structure of rings*, Amer. Math. Soc. Colloq. Publ., vol. 37, Amer. Math. Soc., Providence, R.I., 1956. MR **18**, 373.

[46] ———, *Lie algebras*, Interscience Tracts in Pure and Appl. Math., vol. 10, Interscience, New York, 1962. MR **26** #1345.

[47] S. A. Jennings, *The group ring of a class of infinite nilpotent groups*, Canad. J. Math. **7** (1955), 169–187. MR **16**, 899.

[48] L. Kaloujnine, *Über gewisse Beziehungen zwischen einer gruppe und ihren Automorphismen*, Berliner Math. Taguag (1963), 164–172.

[49] M. I. Kargapolov, *Finitary approximability of supersolvable groups with respect to conjugacy*, Algebra i Logika Sem. **6** (1967), no. 1, 63–68. (Russian) MR **35** #6741.

[50] A. I. Kostrikin, *The Burnside problem*, Izv. Akad. Nauk SSSR Ser. Mat. **23** (1959), 3–34; English transl., Amer. Math. Soc. Transl. (2) **36** (1964), 63–99. MR **24** #A1947.

[51] W. Krull, *Jacobsonsches Radikal und Hilbertscher Nullstellensatz*, Proc. Internat. Congress Math. (Cambridge, Mass., 1950), vol. 2, Amer. Math. Soc., Providence, R. I., 1952, pp. 56–64. MR **13**, 526.

[52] A. G. Kuroš, *The theory of groups*, GITTL, Moscow, 1953; English transl., Vols. I, II, Chelsea, New York, 1955, 1956. MR **15**, 501; MR **17**, 124.

[53] M. Lazard, *Sur les groupes nilpotents et les anneaux de Lie*, Ann. Sci. École Norm. Sup. (3) **71** (1954), 101–190. MR **19**, 529.

[54] A. Learner, *Residual properties of polycyclic groups*, Illinois J. Math. **8** (1964), 536–542. MR **29** #3532.

[55] F. Levin, *Generating groups for nilpotent varieties*, J. Austral. Math. Soc. Soc. **11** (1970), 28–32.

[56] J. Lewin, *A finitely presented group whose group of automorphisms is infinitely generated*, J. London Math. Soc. **42** (1967), 610–613. MR **36** #5200.

[57] F. Levi and B. L. van der Waerden, *Über eine besondere Klasse von Gruppen*, Abh. Math. Sem. Univ. Hamburg. **9** (1933), 154–158.

[58] S. MacLane, *Homology,* Die Grundlehren der math. Wissenschaften, Band 114, Academic Press, New York; Springer-Verlag, Berlin, 1963. MR 28 #122.

[59] W. Magnus, *Bexiehungen zwischen Gruppen und Idealen in einem speziellen Ring,* Math. Ann. **111** (1935), 259–280.

[60] ———, *Über freie Faktorgruppen und freie Untergruppen gegebener Gruppen,* Monatsh. Math. Phys. 47 (1939), 307–313.

[61] W. Magnus, A. Karrass and D. Solitar, *Combinatorial group theory: Presentations of groups in terms of generators and relations,* Pure and Appl. Math., vol. 13, Interscience, New York, 1966. MR **34** #7617.

[62] A. I. Mal'cev, *On certain classes of infinite solvable groups,* Mat. Sb. **28** (70) (1951), 567–588; English transl., Amer. Math. Soc. Transl. (2) **2** (1956), 1–21. MR **13**, 203.

[63] ———, *On a class of homogeneous spaces,* Izv. Akad. Nauk SSSR Ser. Mat. **13** (1949), 9–32; English transl., Amer. Math. Soc. Transl. (1) 9 (1962), 276–307. MR 10, 507.

[64] ———, *Nilpotent torsion-free groups,* Izv. Akad. Nauk SSSR Ser. Mat. **13** (1949), 201–212. (Russian) MR **10**, 507.

[65] ———, *On the faithful representation of infinite groups by matrices,* Mat. Sb. **8** (**50**) (1940), 405–422; English transl., Amer. Math. Soc. Transl. (2) **45** (1965), 1–18. MR 2, 216.

[66] J. C. C. McKinsey, *The decision problem for some classes of sentences without quantifiers,* J. Symbolic Logic **8** (1943), 61–76. MR **5**, 85.

[67] J. W. Milnor, *A note on curvature and fundamental group,* J. Differential Geometry 2 (1968), 1–7. MR 38 #636.

[68] ———, *Growth of finitely generated solvable groups,* J. Differential Geometry 2 (1968), 447–449. MR **39** #6212.

[69] C. C. Moore, *Decomposition of unitary representations defined by discrete subgroups of nilpotent groups,* Ann. of Math. (2) **82** (1965), 146–182. MR **31** #5928.

[70] A. W. Mostowski, *Computational algorithms for deciding some problems for nilpotent groups,* Fund. Math. 59 (1966), 137–152. MR 37 #293.

[71] ———, *On the decidability of some problems in special classes of groups,* Fund. Math. 59 (1966), 123–135. MR 37 #292.

[72] B. H. Neumann, *An embedding theorem for algebraic systems,* Proc. London Math. Soc. (3) **4** (1954), 138–153. MR **17**, 448.

[73] ———, *Ascending derived series,* Compositio Math. 13 (1956), 47–64. MR **19**, 632.

[74] H. Neumann, *Varieties of groups*, Ergebnisse der Mathematik und ihrer Grenzgebiete, Band 37, Springer-Verlag, New York, 1967. MR 35 #6734.

[75] J. Nielsen, *Om Regning med ikke-kommutative Faktoren og dens Anvendelse i Gruppenteorien*, Mat. Tidsskr. B (1921), 77–94.

[76] I. B. S. Passi, *Dimension subgroups*, J. Algebra 9 (1968), 152–182. MR 38 #242.

[77] V. P. Platonov, *Frattini subgroup of linear groups and finitary approximability*, Dokl. Akad. Nauk SSSR 171 (1966), 798–801 = Soviet Math. Dokl. 7 (1966), 1557–1560. MR 35 #2971.

[78] M. O. Rabin, *Recursive unsolvability of group theoretic problems*, Ann. of Math. (2) 67 (1958), 172–194. MR 22 #1611.

[79] V. N. Remeslennikov, *Conjugacy of subgroups in nilpotent groups*, Algebra i Logika Sem. 6 (1967), no. 2, 61–76. (Russian) MR 36 #1545.

[80] D. J. S. Robinson, *A property of the lower central series of a group*, Math. Z. **107** (1968), 225–231. MR 38 #3355.

[81] R. D. Schafer, *An introduction to nonassociative algebras*, Pure and Appl. Math., vol. 22, Academic Press, New York, 1966. MR 35 #1643.

[82] J-P. Serre, *Lie algebras and Lie groups*, Benjamin, New York, 1965. MR 36 #1582.

[83] R. G. Swan, *Representations of polycyclic groups*, Proc. Amer. Math. Soc. 18 (1967), 573–574. MR 35 #4306.

[84] B. A. F. Wehrfritz, *Sylow theorems for periodic linear groups*, Proc. London Math. Soc. (3) 18 (1968), 125–140. MR 36 #3893.

[85] ———, *Soluble periodic linear groups*, Proc. London Math. Soc. (3) 18 (1968), 141–157. MR 36 #3894.

[86] J. A. Wolf, *Spaces of constant curvature*, McGraw-Hill, New York, 1967. MR 36 #829.

[87] ———, *Growth of finitely generated solvable groups and curvature of Riemannian manifolds*, J. Differential Geometry 2 (1968), 421–446.

[88] H. Zassenhaus, *Über Lie'sche Ringe mit Primzahlcharakerteristik*, Abh. Math. Sem. Univ. Hamburg. 13 (1940), 1–100.

[89] M. Zorn, *On the theorem of Engel*, Bull. Amer. Math. Soc. 43 (1937), 401–404.

John Oprea
101 Carl Dr.
Columbus, Ohio
43210